AF809050

Field Manual
No. 1-564

***FM1-564**
Headquarters
Department of the Army
Washington, DC

Contents

PREFACE

CHAPTER 1. PREDEPLOYMENT PLANNING

Section 1. Mission Analysis

1-1.	Preparation	1-1
1-2.	Mission Definition	1-1
1-3.	Shipboard Helicopter Training Requirements	1-2
1-4.	Service Responsibilities	1-2
1-5.	Logistics	1-3

Section 2. Presail Conference

1-6.	Coordination	1-7
1-7.	Number of Army Aircraft on Board the Ship	1-7
1-8.	Checklist	1-7

Section 3. Training Requirements

1-9.	Aircrew Requirements for Training	1-9
1-10.	Ground School Training	1-11
1-11.	Initial Qualification and Currency Requirements	1-11
1-12.	Ship Certification and Waiver	1-15
1-13.	Detachment Certification	1-15

CHAPTER 2. PREPARATION FOR FLIGHT OPERATIONS

Section 1. Chain of Command

2-1.	Command Relationship	2-1
2-2.	Special Operations	2-2
2-3.	Augmentation Support	2-2

Section 2. Personnel Responsibilities

2-4.	Flight Quarters Stations	2-3
2-5.	Landing Signal Enlisted	2-4

DISTRIBUTION RESTRICTION: Approved for public release; distribution is unlimited.

Section 3. Aircraft Handling

2-6.	Fundamentals	2-4
2-7.	Helicopter Recovery Tie-Down Procedures	2-5

Section 4. The Air Plan

2-8.	Scope	2-5
2-9.	Contents	2-6
2-10.	Maintenance Test Flights	2-7
2-11.	Flight Plan	2-7
2-12.	Aqueous Film-Forming Foam System and Mobile Firefighting Equipment	2-7

CHAPTER 3. SHIPBOARD AIR TRAFFIC CONTROL

3-1.	Responsibilities	3-1
3-2.	Aircraft Control Criteria	3-1
3-3.	Control Zone or Control Area Limitations	3-4
3-4.	Aircraft Separation Criteria	3-4
3-5.	Communications Control	3-5
3-6.	Emergency Control Procedures	3-6
3-7.	Transient Aircraft	3-7
3-8.	Lost Aircraft Procedures	3-7

CHAPTER 4. LAUNCHING AIRCRAFT

Section 1. General Information

4-1.	Operational Procedures	4-1
4-2.	Air Officer and Aviation Unit Operations Duty Officer Responsibilities	4-4

Section 2. Flight Deck Procedures

4-3.	General Procedures	4-4
4-4.	Preflight Inspection	4-11
4-5.	Prelaunch Procedures	4-12

Section 3. Aircraft Launches

4-6.	Launch Procedures	4-16
4-7.	Night Launches	4-17
4-8.	Night Vision Device Considerations	4-18
4-9.	EMCON or Zip-Lip Launch Procedures	4-20
4-10.	Emergency After Launch	4-21
4-11.	Helicopter Departure Procedures	4-22
4-12.	Control of Departing Aircraft	4-22

CHAPTER 5. AIRCRAFT RECOVERY

	Section 1. Arrival	
5-1.	Procedures	5-1
5-2.	Lost Communications or Navigations Aids During the Approach	5-11
5-3.	Aircraft Diversion	5-13
	Section 2. Recovery	
5-4.	Preparation	5-15
5-5.	Recovery With Ordnance	5-19
5-6.	Helicopter Recovery Tie-Down Procedures	5-20
5-7.	Night Recovery	5-21
	Section 3. Safety	
5-8.	Special Safety Precautions	5-25
5-8.	Emission Control, Zip-Lip, and Lost Communication Procedures	5-25
CHAPTER 6.	**AIRCRAFT AND WEAPONS HANDLING**	
	Section 1. Overview	
6-1.	General Requirements	6-1
6-2.	Safety Briefing	6-1
6-3.	Maintenance Liaison Officer	6-1
6-4.	Equipment	6-2
	Section 2. Aircraft Handling	
6-5.	Aircraft Movement	6-2
6-6.	Fueling and Defueling Aircraft	6-7
6-7.	Medical Casualties on the Flight Deck	6-9
	Section 3. Weapons Handling	
6-8.	Procedures	6-9
6-9.	Hazards of Electromagnetic Radiation to Ordnance and Safety Precautions	6-10
6-10.	Weapons Movement and Handling	6-10
6-11.	LHA Class Weapons Handling Restrictions	6-12
6-12.	Weapons Assembly and Disassembly	6-12
6-13.	Weapons Loading and Downloading	6-12
6-14.	Arming	6-14
6-15.	Dearming	6-14
6-16.	Abort Strikedown	6-15
6-17.	Maintenance on Loaded Aircraft	6-15

APPENDIX A. AIRCRAFT HANDLING SIGNALS A-1

APPENDIX B. AIRCRAFT ARMING AND SAFING SIGNALS B-1

APPENDIX C. WEAPONS LOADING, STRIKEDOWN, DOWNLOADING, AND RECOVERY GUIDE C-1

APPENDIX D. OPERATIONS FROM SINGLE-AND DUAL-SPOT SHIPS D-1

APPENDIX E. STANDING OPERATIONS PROCEDURES FOR OVERWATER OPERATIONS E-1

APPENDIX F. FLIGHT DECK CLOTHING AND DUTIES F-1

APPENDIX G. HELICOPTER/SHIP INTERFACE G-1

APPENDIX H. MEMORANDUM OF UNDERSTANDING H-1

GLOSSARY Glossary-1

REFERENCES References-1

PREFACE

This manual describes the tactics, techniques, and procedures for use by Army aviation units during operations from Navy and Coast Guard ships. It is written to reflect peacetime operations that may transition into warfighting execution and assumes that the deployment of Army helicopters is the result of careful presail planning.

This manual is intended for commanders, staffs, aircrews, and instructors. It will be used to coordinate, plan, execute, and teach shipboard operations. Along with Navy publications, it provides information for developing a standardized, progressive program to train crews to proficiency on shipboard operations. Appendixes A through F provide supplemental information on aircraft handling signals; aircraft arming and safing signals; weapons loading, strikedown, downloading, and recovery guide; operations from single- and dual-spot ships; standing operating procedures for overwater operations; and flight deck clothing and duties. Appendix G provides information on helicopter/ship interface; the most current memorandum of understanding between the Army, Air Force, and Navy for deck landing operations is found in Appendix H. This publication also reflects Navy terminology, regulations, procedures, and traditions that are necessary for safe operation aboard ships. Unless stated otherwise in the text, the term "battalion" refers to both aviation battalions and squadrons.

The proponent of this publication is HQ TRADOC. Aviation units are encouraged to recommend ideas to improve the tactics, techniques, and procedures in this manual. Send comments and recommendations on DA Form 2028 (Recommended Changes to Publications and Blank Forms) directly to the Commander, US Army Aviation Center, ATTN: ATZQ-TDS-DM, Fort Rucker, AL 36362.

Unless stated otherwise, masculine nouns and pronoun do not refer exclusively to men.

This manual has been reviewed for operations security considerations.

CHAPTER 1

PREDEPLOYMENT PLANNING

In nearly every major conflict and operation since World War II, Army aviation has been assigned missions in the maritime environment, either basing off naval vessels for land attack or operating from ships for sustained overwater missions. In recent years, the nature and complexity of those missions have changed dramatically, dictating that aviation units complete specialized preparatory and sustainment training. Recent worldwide deployments have shown that Army aviation has a versatile combination of equipment sophistication, deployability, and personnel to accomplish specific strategic missions that require operations in the maritime environment.

Section I. Mission Analysis

1-1. PREPARATION

a. Army aviation units are presently participating in many joint operations that require proficiency in shipboard operations to perform--

- Medical evacuation from shore to ship.
- Logistics transfer and resupply.
- Armed and unarmed reconnaissance and sealane surveillance.
- Maritime security operations, small boat interdiction, ship takedown and area denial.
- Attack helicopter operations.

b. FM 100-5 stresses the need for training and preparing for shipboard operations. Chapter 4 states that a force projection army requires extraordinary flexibility in thinking about operations because of the variety of combinations of joint forces available and the range of possible circumstances for their employment. It also states that Army doctrine stresses unified air, land, sea, and special operations--all supported by space operations--throughout the theater of war. This publication helps planners prepare for air-sea missions, specifically those missions that require landing on and operating from US Navy and Coast Guard air-capable ships.

c. The document that governs Army shipboard operations is the "Army/Air Force Deck Landing Operations Memorandum of Understanding" signed by the Army, Air Force, and Navy in July 1988. Information from this MOU will be supplemented as necessary to provide more comprehensive guidance in planning and conducting Army aviation shipboard flight operations.

1-2. MISSION DEFINITION

Shipboard or overwater specified tasks are found in nearly all regional contingency plans, JTF plans, and counternarcotics operations. Shipboard missions require deck landings and support operations performed from a ship. Overwater missions include operations over open water but originate and/or end at a land base. Therefore, units must precisely define their missions. Scarce aviation and naval resources dictate that shipboard operations be a priority mission-essential task for a unit to conduct this training. The priority of the mission is determined by higher command or by the emergent nature of the training requirement.

1-3. SHIPBOARD HELICOPTER TRAINING REQUESTS

a. ***Emergency Training.*** Emergency requirements or training necessary for imminent deployment usually will come from DA through a MACOM such as FORSCOM. Army aviation units will receive the higher command's assistance in scheduling ships and other resources.

b. ***Extended Training Requirements.*** Shipboard training requirements that result from mission analysis of contingency plans, directives, or open-ended commitments follow an established procedure to schedule ships and other resources. This procedure is discussed below.

(1) Each unit that wants to conduct shipboard training must submit an annual deck requirements forecast for the next fiscal year through its MACOM chain of command.

NOTE: Units should contact their MACOM for the specific format for this request.

(2) After receiving MACOM approval, units must submit back through their MACOMs the detailed quarterly scheduling requests. The scheduling requests are consolidated and submitted to the appropriate fleet commander's staff and scheduling conference. The fleet commander's staff deals with routine scheduling requests two quarters in advance; service requests must arrive 45 days before the scheduling conference. The fleet distributes suspense dates in message format.

(3) The Army force representative to the naval surface command or the aviation staff officer from the MACOM attends the scheduling conference to ensure that Army service requests are filled properly. The Army force representative also tracks each request and assists with schedule changes, presail conferences, and other coordination.

c. ***Overwater Training.*** Overwater training involves environmental factors which are inherent during shipboard operations. Units must plan and train for safe, extended overwater flight. An example of an overwater operations SOP is in Appendix E.

1-4. SERVICE RESPONSIBILITIES

The joint force MOU identifies service responsibilities that support shipboard operations. These responsibilities are discussed below.

a. ***Navy.*** The fleet commander's staff will schedule Army requests for DLQ services on ships that are staffed with personnel who are certified to conduct shipboard training and/or overwater gunnery.

(1) The Navy can make specific personnel available for ground school or flight operations training, including HACs and LSEs. The Army must provide the helicopters for the training and is responsible for helicopter operating costs. In addition, TDY costs for Navy personnel who provide DLQ training for Army aviators are paid by the Army. When travel is required, funding information must be provided before TDY orders are cut.

(2) The Navy can furnish the necessary publications to complete reference libraries. Units can use normal requesting procedures to obtain these publications.

(3) Army unit training requests that were defined in the presail conference will be complied with according to Naval regulations and at the discretion of the naval air-capable ship commander.

(4) Qualified personnel on board naval air-capable ships will conduct deck-landing services for Army units that are embarked or are operating from shore.

b. ***Army Aviation.*** Commanders of units scheduled to conduct DLQ training will ensure that training and logistics prerequisites for shipboard helicopter operations are satisfied. Training units must brief all planned missions at the presail conference to ensure that the ship can safely comply with all requests. Army units must understand and follow naval aviation and shipboard regulations in all instances.

c. ***Overwater Gunnery Training.*** Risk management must be performed for all training events. During the unit's handling and employment of explosive ordnance, commanders must be involved in implementing proper countermeasures for safe mission accomplishment. Aviation units wanting to conduct overwater gunnery while basing from a ship must identify all ordnance intended for onboard stowage during the presail conference. All ordnance must be certified shipboard-safe according to NAVSEA OP-4 before it is loaded.

d. ***Night Vision Device Training.*** Shipboard NVD operations are authorized by the Chief of Naval Operations, Code N889. On an annual basis, units must request authority to train with NVDs through their chain of command to DAMO-TRO, ODCSOPS.

e. ***Nonstandard Aviation Maneuver Training.*** The ship commander obtains the necessary guidance and regulatory information to approve or modify training plans. Therefore, any request to conduct nonstandard aviation maneuvers in the shipboard environment must be briefed thoroughly at the presail conference.

1-5. LOGISTICS

The logistics requirements to support a shipboard training service depend on the length of the service and whether the unit embarks or maintains a shore base.

a. ***Publications.***

(1) Army units can use the AG publications system to order joint publications. For information on Navy publications, call the Navy Publications Customer Service Center, Philadelphia, Pennsylvania, DSN 442-2997/2626/0160/2267. For NAVAIR manual distribution information, call the Naval Air Technical Services Facility, DSN 442-4670. To request automatic distribution or request specific manuals, send a memorandum on letterhead stationary to the CO, Naval Air Technical Services Facility, Code 25, 700 Robbins Avenue, Philadelphia, PA 9111-5097. The memorandum, signed by the unit commander, will contain the following information:

(a) A request for automatic distribution along with initial issues of manuals. Provide stock numbers of manuals, which are included in the References section of this publication; the number of copies needed for initial distribution; and the number of copies needed for automatic distribution. For example, I request 10 copies be sent to this unit when the new manual is distributed.

(b) The unit designation, address, UIC, and the type of aircraft in the unit.

(c) A point of contact, DSN, and commercial phone number.

(2) For unclassified NWP distribution and confirmation information, contact the Navy Tactical Support Activity at DSN 288-6163 before writing. After confirmation, send a memorandum on letterhead stationary through the Commander, Naval Doctrine Command, 1540 Gilbert Street, Norfolk, VA 23511-2785, to the Director, Navy Tactical Support Activity, 901 M Street SE, Building 200 (Code 57), Washington, DC 20374-5079. The memorandum, signed by the unit commander, will contain the following information:

(a) A request for automatic distribution, along with initial issues of NWP manuals. Provide stock numbers of manuals, the number requested for initial distribution, and the number requested for automatic distribution. For example, I request 10 copies be sent to this unit when the new manual is distributed.

(b) The unit designation, address, UIC, and the type of aircraft in the unit.

(c) A point of contact, DSN, and commercial phone number.

b. ***Funding.*** Deck services are conducted on the ship along with the normal training schedule. Therefore, the Army does not incur use or fuel charges. The S4 of the embarking unit must provide accounting data to the ship supply officer to pay for supplies, fuel, and parts not embarked with the unit. Meals and berthing are paid through an independent fund.

c. ***Fuel.***

(1) All naval aircraft use JP5 fuel only. Army units must ensure that their aircraft use JP5 before embarkation. The fuel system must not contain residual JP4 or JP8. If flashpoint readings from the fuel sample are not within tolerance (below 120 degrees Fahrenheit), the aircraft cannot be hangared on the ship nor can the aircraft be brought to the hangar deck. If JP5 cannot be used before embarkation, the preferred method of switching from JP4 or JP8 to JP5 is for the aircraft to arrive at the point of embarkation with only enough fuel left to complete loading on the deck. Then JP5 can be pumped into the aircraft, diluting the remaining fuel and meeting the flashpoint requirement.

(2) If removed from the aircraft, the UH-60 ESSS or AH-64 external fuel tanks must be purged before they can be taken to the hangar deck. This requirement must be planned for because of delays and possible storage space problems.

d. ***Refueling.*** Navy standards for hot refueling are absolute: "zero tolerance" for leaks, seeps, or dripping during refueling.

(1) Units that have unique refueling devices, such as Wiggins nozzles, must bring them on board during their deck service period to ensure proper refueling procedures. To ensure trouble-free service, Wiggins nozzles should be pressurized and checked before deployment.

(2) The compatibility of these nozzles on Navy vessels vary from ship to ship because of the different type of connectors available on the ships. A Wiggins or other refueling device should be taken to the presail conference so that the ship's fuels officer can check its compatibility with Navy equipment.

e. *Reimbursement*. The S4 of the unit must have proper accounting code information from the unit comptroller or budget analyst to pay for fuel. Reimbursement for fuel is requested using DD Form 1348 (DOD Single Line Item Requisition System Document); identiplates usually are not accepted.

f. *Assignments of Embarked Units.*

(1) When an Army unit embarks on the Navy ship, it can expect to be assigned duties on the ship to augment the ship's company. These duties may include security watch, cleaning details, fire watch, and mess support. All assignments will be made through the Army chain of command on the ship.

(2) The embarked OH-58D(I) detachment may not be required to accept additional assignments because of the small number of personnel assigned to the detachment. These detachments normally require that all deployed personnel conduct flight operations; therefore, all personnel must be on the same duty cycle.

g. *Meals.*

(1) Officers may eat their meals in the officer's wardroom for a nominal daily fee that covers berthing. The uniform is generally the duty uniform. Senior enlisted members eat in the chief petty officer's dining room for a nominal daily fee, which also covers berthing. Enlisted soldiers (E-6 and below) eat in the ship's mess and are berthed without charge. Because the Navy handles shipboard meals differently from field rations, the unit must ensure that meals are paid for before disembarking from the ship.

(2) In the ward room, Navy officers wear the equivalent of the class B uniform. Flight suits normally are not allowed in the ward room. The issuance of proper attire for Army officers on the ship should be addressed during the presail conference.

h. *Aircraft Maintenance.*

(1) Support facilities vary from ship to ship. (The Aviation Facilities Resume should be consulted for specific capabilities.) Units must plan to support their unique aircraft maintenance requirements by preparing logistics replacement units and pack-up kits for deployment to the shore base or to the ship.

(2) Army aircraft are not manufactured to the anticorrosion standards of Navy aircraft and are prone to corrosion. Units should plan to purchase an anticorrosion compound for their aircraft before embarkation. Recent experience has shown that unprotected major aircraft components lose an estimated 25 to 30 percent of their useful life because of saltwater corrosion.

(3) Freshwater washes can be conducted on board. However, large numbers of aircraft and GSE on the deck may prevent this from being feasible.

i. ***Ground Support Equipment.*** Embarking units must identify all equipment being brought on board so that safety and compatibility checks can be made. Every effort should be made to ensure that the embarked Army aircraft and the ship's GSE are compatible. Very limited space exists for extra GSE.

(1) ***Ground-handling wheels and tow bars.*** Units must learn to position aircraft on deck and in the hangar bays. Multiple ground-handling wheels and tow bars are essential for rapid movement of aircraft such as during a fire-fighting sequence.

(2) ***Blade folding kits.*** Aircraft modified with a blade folding capability must deploy with the proper blade folding kit to allow movement into hangars. In addition, blade struts also must deploy with aircraft. When these aircraft are positioned on the flight deck, they are vulnerable to damage when the blades flap in the wind. Blade struts or some other device to secure the blades will give the aircraft added protection.

j. ***Aviation Life Support Equipment.*** (ALSE requirements are according to AR 95-3.)

(1) Additional ALSE that is needed before embarkation includes HEEDS bottles for all aircrew members. Life rafts are required for overwater operations in UH-1, UH-60, and CH-47 aircraft..

(2) All aircrew members must have an LPU (life preserver) for overwater operations. According to AR 95-1 and AR 95-3, LPUs also are required for soldiers being transported in UH-1, UH-60, and CH-47 aircraft. Adequate LPUs must be identified by type and NSN during planning. Currently, LPU-21-23 series are adequate for fully equipped combat soldiers of all possible weights and sizes. The LPU-10 series is inadequate, and presents a significant safety hazard. The ARSOA standard is the LPU-21/P, NSN 4220-00-220-4894.

(3) The ship has very limited ALSE for passengers. Every effort should be made to obtain extra ALSE for any passengers that units may have to transport. During night operations, all flight deck personnel must wear a flotation device; this includes crews performing maintenance on the flight deck.

(4) Coordination with the Navy must be effected to secure all of the required ALSE before embarkation. In addition, survival vests should be upgraded with additional sun and water protection items.

k. ***Mail.*** Mail service is available for moderate-sized parts, packages, personal, and official mail. The ship's address must be confirmed at the presail conference and sent to the whole unit before embarkation.

l. ***Message Address.*** Navy message traffic is sent by message text format. Accordingly, Army units must become accustomed to message format procedures. During the presail conference, Army units must provide the appropriate personnel on the ship with the message addresses of the training unit and the next higher command.

m. ***Customs and Courtesies.*** The Navy is rich in customs and courtesies, particularly when the ship is underway. All soldiers in the embarking unit must understand the unique traditions of the Navy and how to respond to them. During the presail conference, the Army unit must confirm with the Navy representative the customs and courtesies that differ from the Army and the ones that must be emphasized during preembarkation training. This may include such differences as rank (a Navy O3 is a lieutenant) and also the procedures for boarding and departing a ship.

Section II. Presail Conference

1-6. COORDINATION

a. The presail conference takes place on board the host ship. This conference is the coordination meeting between the host ship and the operating unit for safety and operational planning.

b. Once a ship has been assigned to fill a specific Army request, the Navy will approve direct coordination (DIRLAUTH) so that the aviation unit can coordinate with the ship for an acceptable presail conference date. The Army force representative also can help in setting a date for the conference, which generally occurs as close as possible to the actual service date.

c. Currently, every deployed amphibious group has an Army officer (normally an Aviation Branch major) assigned as a liaison officer. Army aviators also are authorized as liaison officers with TACRONS. These officers are very important in coordinating aviation-specific logistical and training requirements with the Navy.

1-7. NUMBER OF ARMY AIRCRAFT ON BOARD THE SHIP

a. During the presail conference, planners decide on the number of Army aircraft that can be brought on board the ship. The implication is that when an Army unit is given the mission to embark on a Navy ship, the assigned ship will be able to carry the unit's aircraft or the number of aircraft required by the directive. When operating in the shipboard environment, aircraft loading involves planning in three areas: These areas are the numbers of aircraft in traffic, on the flight deck; and/or in the hangar bay or hangar deck.

b. The Navy determines the number of aircraft that can be deployed on the assigned ship. The following considerations may be included in this decision.

(1) If an air-capable ship is assigned that has no specific Army helicopter interoperability data and the helicopters are smaller than a CH-47, the CH-53 operating radius for space analysis may be used. However, landing certification is primarily based on structural data of the deck. Usually, if the deck can hold a CH-53, it can hold a CH-47 or an AH-64. As a guide, the matrices in Appendix G show the specifics of interoperability between Army aircraft and Navy ships.

(2) A large aircraft carrier like the USS Eisenhower (CVN 69) could carry hundreds of Army helicopters. However, operations, including VERTREP, must continue. This precludes stacking helicopters on every inch of the flight deck, which becomes more important on smaller, single- or dual-spot ships. The Navy will consider its own missions, as well as flight-deck mobility and fire-fighting and/or rescue capabilities, when planning for

Army aircraft. Because the size of the aircraft cannot be quickly reduced and moved off of landing spots, Army units being trained normally will operate with only as many aircraft as can be landed on the flight deck.

NOTE: A flight deck covered with Army helicopters may slow or prevent a timely rescue if an emergency should occur on the ship during flight operations.

(3) During deployment, the aircraft may be parked with their tailbooms hanging off of the flight deck out over the water. The AH-64 and UH-60 cannot be parked this way because their landing gear is installed at the extreme end of the tailboom. Therefore, a unit equipped with UH-60 or AH-64 aircraft may not be able to carry as many aircraft on board a ship as an OH-58D(I) or a UH-1 unit.

c. When the ship is identified and the number of Army aircraft to be deployed is decided, Army planners may gain specific information on the ship from the Navy manual NAEC-ENG-7576. This manual contains diagrams of all Navy ships and their landing facilities. Additional questions may be addressed by calling the Shipboard Aviation Facility Hot Line. This facility can confirm--

- Criteria or standards specified in Air-Capable Ship Aviation Facilities Bulletins.
- Criteria or standards specified in Amphibious Assault Ship Aviation Facilities Bulletins.
- Criteria or standards specified in Visual Landing Aids General Service Bulletins.
- Ship certification status.
- Shipboard equipment, configuration, and deficiencies.
- Any other matter relative to aviation facilities aboard air-capable and amphibious aviation ships.

The hot line phone number is DSN 624-2592 or commercial (908) 323-2592. Written correspondence should be addressed to--

Head, Performance and Certification Branch

Support Equipment and Aircraft Launch and Recovery Equipment

In-Service Engineering Division (Code 4.8.10.4)

Naval Aviation Warfare Center

Aircraft Division

Highway 547

Lakehurst, NJ 08733-5000

1-8. CHECKLIST

Figure 1-1 (page 1-12) is a sample format for a checklist that may be used to cover all the necessary coordination topics during the presail conference. Units are encouraged to modify this sample to meet specific unit requirements. In addition to information on the checklist, aviation units will provide diagrams of assigned aircraft showing aircraft egress, fuel cell locations, tie-down points, and desired wind envelopes.

Section III. Training Requirements

1-9. AIRCREW REQUIREMENTS FOR TRAINING

The requirements in this section are based on the Army-Navy-Air Force MOU dated 1988. The MOU and its requirements may change. Therefore, before the units complete any training, they should contact their MACOMs to confirm training requirements. The current requirements are shown below.

- Army aviators must be qualified and current according to AR 95-1.
- PCs will be deck-landing qualified and current.
- Pilots performing deck landings will be deck-landing qualified and current unless they are undergoing training.
- Flight training must be conducted by an approved Navy HAC or an Army IP or UT who is deck-landing qualified and current in the aircraft and in the flight mode. Units just beginning their DLQ program require a current IP from another unit to qualify the new IPs.

SHIPBOARD HELICOPTER OPERATIONS

PRESAIL CONFERENCE CHECKLIST

	UNIT:	POC:	DSN:
AVIATION:	____________	____________	________
SHIP:	____________	____________	________

1. Establish
   a. DLQ date ________
   b. Flight schedule ________
2. Field Deck-Landing Qualification Requirements (Ref: Army-Air Force-Navy MOU) ________
3. DLQ Currency Requirements (Ref: Army-Air Force-Navy MOU) ________
4. Type and Number Aircraft Involved ________
5. Pilots Needing Initial Qualification/ Currency (Ref: Army-Air Force-Navy MOU) ________
6. Surface/Air Clearances ________

(Ship Responsibility)

7. Aviation Facility Waiver (Type Commander Will Coordinate) ____________

8. Transient A/C Local Ops Briefs (Base Ops Provides) ____________

9. TACAN/Radio Frequencies ____________

10. Ship Overhead Msg (Containing Ops/Comm Info) ____________

11. Safety/Operations Brief) (Ship/NWP 3-04.1M) ____________

12. Crash Rescue Procedures and Postcrash Fire Procedures ____________

13. Search and Rescue ____________

14. Ships Glideslope Indicator (Different from Army GSI) ____________

15. Engage/Disengage Envelopes (Shipboard) ____________

16. No-Rotor Brakes on Some Army Helicopters ____________

17. Takeoff/Recovery Envelopes ____________

18. Bad Weather Procedures ____________

19. Fuel Requirements on Board Ships

a. JP5 only ____________

b. NATO 01 or Wiggins Nozzles ____________

20. Fuel Reimbursement (Standard Military Credit Card or DD Form 1348 ____________

21. Name of Army/Air Force Liaison Officer (During Shipboard DLQ Period) ____________

22. Shore-Based Administrative/ Logistics Coordinator ____________

a. Helicopter Ramp Parking* ____________

b. Freshwater Wash ____________

c. Accommodations:

Officer ____________

Enlisted	________
d. Mess Facilities	________
e. Local Transportation	________

*Contact base air operations for transient parking and to obtain POC phone numbers for other logistical requirements.

COMNAVSURFLANT POC: Aviation Directorate, Operations Officer, Norfolk, VA, commercial (804) 444-8623/8633 or DSN 564-8623/8633.

COMNAVSURFPAC POC: Aviation Directorate, Operations Officer, San Diego, CA, commercial (619) 437-2311 or DSN 577-2311.

Figure 1-1. Sample format for a presail conference checklist

1-10. GROUND SCHOOL TRAINING

Aircrews must receive instruction to become familiar with the mandatory operational procedures and training requirements for shipboard helicopter operations. The ground school course will include but not be limited to--

- Aircraft landing and handling signals.
- Deck markings and lighting orientation.
- Emergency procedures.
- Communications, NAVAIDs, and EMCON.
- Fuel and/or maintenance support and procedures.
- Landing patterns and/or approaches and ship control zones.
- VERTREP procedures, if applicable.
- Presail conference procedures.

1-11. INITIAL QUALIFICATION AND CURRENCY REQUIREMENTS

a. ***Single-/dual-spot ships.*** Initial qualification and currency requirements are outlined in the following paragraphs.

NOTE: Hereafter single-/dual-spot ships will be referred to as single-spot ships.

(1) ***Initial day qualification,***

(a) Flight training is conducted by a US Army or USAF deck-landing qualification SP or by a US Navy HAC who is current on single-spot decks. Army IPs or SPs must conduct qualification of Army aviators.

(b) Ground school training is conducted according to paragraph 1-10.

(c) Six field deck landings must be conducted before six single-spot shipboard landings (all within a ten consecutive day period).

(2) ***Currency requirements***: Four single-spot shipboard landings must be conducted within 90 days.

(a) Pilots whose currency has lapsed but have made four single-spot landings within the last 180 days will--

- Undergo training conducted by either a current DLQ PC or DLQ IP.
- Perform four field deck landings before six shipboard landings (all within a ten-consecutive-day period).

(b) Pilots whose currency has lapsed and who have not made four single-spot landings within the last 181 days must undergo initial qualification training.

NOTE: Night single-spot helicopter operations require more training and more specialized equipment than day operations. Some units, particularly OH-58D(I) and MEDEVAC detachments, are required to perform these operations during the normal conduct of their missions. Requests for this type of training will be handled by the US Navy (OP-593) and the US Army (DAMO-TRS). Appendix D contains more information on operations from single-spot ships.

b. ***Multispot Ships (LPH/LHA/CV).***

(1) ***Initial day qualification.*** Initial day qualification requirements are outlined below.

(a) Flight training is conducted by a US Army or US Air Force IP or UT who is day-current.

(b) Ground school training is conducted according to paragraph 1-10.

(c) Five day field deck landings must be conducted before five day shipboard landings (all within a ten consecutive day period).

(2) ***Day currency requirements***: Four shipboard landings must have been made within the preceding nine months. Pilots whose day currency has lapsed will undergo initial day qualification; requalification will be conducted by an Army IP, UT, or PC.

(3) ***Initial night qualification.*** Initial night qualification requirements are outlined below.

(a) The pilot must be day-qualified and current.

(b) Ground school training is conducted according to paragraph 1-10.

(c) Flight training will be conducted by a night-current US Army DLQ SP or US Navy HAC.

(d) Six night field deck landings must be conducted before six night-shipboard landings (all within a ten consecutive day period). Pilots also must comply with the 72-hour requirement in paragraph (4) below.

NOTE: Some Navy ships, particularly CV/CVNs, do not have NVG-compatible deck lighting. As a result, the intensity level of the flight deck lighting must be lowered during NVG operations. Normally, the ship does not have NVG on board. The Army detachment should be prepared to provide primary flight control with at least two sets of NVG and the bridge team with two sets.

(4) ***Night currency requirements***: To maintain currency, six night shipboard landings must have been made within the preceding 90 days. If more than 72 hours have lapsed since the last night shipboard landing, one day shipboard landing will be performed within 24 hours before the next night shipboard landing.

c. ***Single-Spot DLQ Training.*** Single-spot DLQ training is the most demanding because of the size of the deck space and the size of the ship. Pilots qualified on single-spot ships are qualified on multispot ships, but the reverse is not true.

d. ***Aircraft Carriers (CV):*** Routine DLQ training and operations normally will not be conducted on CV class ships. Operations on CV class ships will be on a case-by-case basis and require a special ground briefing by Navy personnel or by Army/Air Force personnel designated by the Navy. Pilots qualified and current on single- and multispot ships are considered qualified and current on CV class ships.

e. ***LOTS and VERTREP Operations.*** Pilots performing LOTS or VERTREP operations that involve external loads without a shipboard landing must be deck-landing qualified and current. Pilots scheduled to participate in LOTS and/or VERTREP operations must receive a familiarization of the designated ship by US Navy personnel, a previously familiarized US Army IP or PC, or a US Air Force IP or FE. (Specific LOTS/VERTREP requirements can be found in NWP 3-04.1.) The familiarization should include--

- Deck markings.
- Cargo staging.
- Communications.
- Load delivery.
- Returning VERTREP equipment and retrograde.
- Staging and pickup of loads and returning.
- Ship lighting.
- Night VERTREP.
- Signaling and communications.

f. ***Techniques for Aircrew Currency.*** Once a unit is identified as requiring continual shipboard currency for its aircrews, the command should accomplish the following:

(1) Identify unit IPs and UTs that must stay current in shipboard operations. Units will make currency a priority mission for these officers.

(2) Continually assess the longevity of the shipboard operations trainers and identify officers with enough time on station to take over training responsibility when needed.

(3) Establish liaison with Navy, Coast Guard, or Marine units that can help keep the unit shipboard operations trainers current. For example, a Marine aviation squadron may be able to include an Army aircraft and crew in their shipboard currency training.

(4) As appropriate, FDLP should be conducted with associated glideslope equipment (non-NVD operations only) and to a spot bearing the appropriate flight deck markings. At a minimum, these provide the opportunity for aviators to practice drift-free takeoffs and landings while establishing deck-marking references. FDLP can be conducted at a unit's home station if field-deck markings are according to the Naval Air Engineering Center specifications.

g. ***Helicopter Landing Trainer.*** The Navy operates a single-spot HLT (IX 514) in Pensacola, Florida, which can be scheduled directly through the operating unit. The HLT is a converted ammunition-carrying ship that simulates an FFG's (Oliver Hazard Perry class frigate) landing platform, allowing Army aircraft to perform DLQ on a ship that is underway. The HLT can be used to qualify and maintain currency in both day and NVD flight modes, is available on an hourly rate, and is paid for with home station training funds. Requests for training on the HLT should be addressed directly to the Operations Officer, 211 South Avenue Suite, NAS Detachment HLT, Pensacola, FL 32508 (DSN 922-8790/8791).

h. ***Underwater Egress Training.***

(1) According to Navy instructions, each aircrew member undergoing overwater and shipboard aircraft training must successfully complete N9 training at an underwater egress trainer (dunker). Units also can conduct N7 training with the HEEDS bottles in the dunker pool. To be considered current in underwater egress, aircrew members must complete this training every three years. Devices are scheduled and operated by the Navy at the following locations:

- Pensacola NAS, FL: DSN 922-2688; FAX 922-3862.
- Cecil Field (Jacksonville), FL: DSN 942-5366/2770; FAX 942-2595.
- Cherry Point, NC: DSN 582-4934/4935; FAX 582-4945.
- Norfolk, VA: DSN 564-1329/3720; FAX 565-9284.
- Miramar NAS (San Diego),CA: DSN 577-4158/4159; FAX 577-6359.
- LaMoure NAS (Near Fresno), CA: DSN 949-3201; FAX 949-3227.

(2) Army units must submit annual forecasts for underwater egress training through their MACOMs to DAMO-TRO for inclusion in the fleet commanders' annual program guidance. Units also may contact the facilities directly to coordinate training periods.

i. ***Other Training Requirements.*** Although they are sometimes waived by the fleet commander, the requirements discussed in the following paragraphs are according to Navy regulations:

(1) ***Firefighting school.*** If the Army unit is embarked as part of the ship's company, all soldiers must attend the Navy firefighting school. Requirements differ based on the length of the cruise and the command relationship. This requirement must be confirmed or denied before embarkation.

(2) ***Class II swimming school.*** All personnel must meet the Navy's class II swimming requirements. Certification is a one-time requirement. As with the firefighting school, this requirement must be confirmed or denied before embarkation.

1-12. SHIP CERTIFICATION AND WAIVER

On an individual basis, fleet commanders (via TYCOM) grant waivers to conduct Army/Air Force helicopter operations.

a. Day VFR shipboard operations may be conducted by Army aviators on Navy ships that have been approved for such operations. Night VFR shipboard operations may be conducted by Army aviators on single- and multispot ships that have been approved for such operations.

b. The Shipboard Aviation Facilities Resume lists all US Navy ships (including CVs); describes and depicts aircraft landing, VERTREP, and hover facilities; flight-deck markings, and lighting arrangements. It also shows the helicopters for which deck certification has been granted.

1-13. DETACHMENT CERTIFICATION

a. Before embarkation, unit commanders or other authority will certify helicopter units for shipboard operations. This certification will ensure that training requirements set forth in this publication have been met. It also certifies that the unit has met parent-service training requirements for the intended mission. Any specific training that has not been completed or any additional training requirements needed after embarkation should be briefed during the presail conference.

b. Before operations, the unit OIC will provide diagrams of embarked aircraft to the HCO or air officer (air boss) and to crash and salvage parties, when requested. As a minimum, these diagrams should show aircraft egress, refueling locations, tie-down points, desired wind envelopes, and pitch and roll limitations.

c. If the air department on the ship has not worked with the training aircraft in a long time or if it has newly assigned personnel, the aircraft should be landed and shut down on the flight deck at the start of training and a detailed walk-around of the aircraft conducted. This helps acquaint LSEs and fire and rescue personnel with the location of fuel ports, fire bottles, release handles, and emergency equipment on the aircraft.

CHAPTER 2

PREPARATION FOR FLIGHT OPERATIONS

This chapter provides the basic information required to prepare for flight operations from a Navy or Coast Guard ship.

Section I. Chain of Command

2-1. COMMAND RELATIONSHIP

The principle governing the command relationship with USMC aviation commands embarked for amphibious operations is contained in NWP 3-02.1 and NWP 5-00.3M. This doctrine is similar to the relationship with Army aviation commands embarked on Navy ships. This section supplements Joint Publication 3-04, which provides guidance for command relationships when helicopter units embark on ships. Overall command and control is discussed in the paragraphs that follow.

NOTE: The words "aviation unit" or "unit" refer to Army aviation units deployed on a ship and is synonymous with the words "battalion" and "squadron."

a. ***Joint Force Commander.*** The relationship between Navy and Army forces during the planning and execution of a joint operation requires a parallel chain of command at all levels of the task force organization. Except during the planning phase, the JFC is responsible for the operation. He exercises that authority over the entire force to ensure success of the operation. Army aviation forces may embark without a specific mission before an initiating directive is received. In this case, the Navy and Army commanders have parallel and equal authority as described in NWP 22 until an initiating directive is received that specifies otherwise.

b. ***Officer in Tactical Command.*** The OTC should ensure that all aviation unit personnel are given ample opportunity to become current and maintain currency in both day and night flight operations.

c. ***Army Aviation Commander/OIC.*** The CO or OIC of an Army helicopter unit reports to the ship's commanding officer while embarked.

d. ***Ship's Commanding Officer.*** Navy regulations set forth the authority of the ship's commanding officer regarding the aircraft embarked in or operating from his ship. The commanding officer of the ship will respect the identity and integrity of embarked aviation units, and--

(1) Give all orders through the chain of command as practicable or as an emergency may dictate.

(2) May require that soldiers perform the duties that their special knowledge and skills enable them to perform when he thinks an emergency exists.

(3) Will ensure that the aviation unit commander has knowledge of any degradation in aviation facilities and certification or deficiencies in training and/or qualified flight quarter's personnel.

(4) Will ensure that the unit has the opportunity to remain current in day and night shipboard landing and launch operations.

(5) Will provide heavy weather protection of aircraft, including hangar space, when available, and comply with aircraft securing procedures in NAVAIR 17-1-537.

(6) Will provide IMA support.

e. ***Aviation Unit Commander.*** The unit commander retains operational authority over and responsibility for aircraft employment and safety of flight operations during all embarked phases of the operation. However, this not impair the authority of the Army task force commander or the ship's commanding officer. To ensure efficient operations, certain actions must be completed and provided to or coordinated with the ship's commander. The aviation unit commander will--

(1) Provide information regarding pilot qualifications and limitations.

(2) Provide a complete list of aircraft being deployed. The list will include aircraft tail numbers, SIF codes, and any configuration peculiarities that will affect handling, ordnance loading, or mission capability.

(3) Furnish aircraft limitations.

(4) Schedule and coordinate aircraft, pilots, and crew men.

(5) Conduct pilot briefings.

(6) Provide maintenance status reports.

(7) Ensure that pilots' day and night shipboard qualifications are current.

(8) Ensure that the applicable heavy weather protective measures are taken as listed in aircraft technical manuals and NAVAIR 17-1-537.

(9) Provide an Army aviation representative to man PriFly and/or AOCC/HDC during flight operations.

2-2. SPECIAL OPERATIONS

a. When command relationships must be modified for special operations, they will be defined in the applicable governing directive, OPLAN, OPORD, or LOI. Normally, units embarked on special operations have the same parallel command relationship as an organization embarked for joint operations. When an aviation unit is directed to embark for a special operation, the CO or OIC of that unit reports to the officer who will conduct the special operation.

b. In some cases, the ship's commander may be assigned as the commander of the special operation. As such, the ship's commander assumes the same posture as a task force commander in his relationship with embarked aviation units. The parallel command relationship of the special operation is maintained. This does not authorize the ship's commander to task nonmission-related flight operations nor does it supersede the inherent aviation unit command responsibilities. Unless the initiating directive states otherwise, aircraft units are under the command of the unit commander or OIC. They are not under OPCON of the ship's commander.

2-3. AUGMENTATION SUPPORT

a. ***Intermediate Maintenance Activity.*** The appropriate service organization will provide augmentation following fleet directives.

b. ***Integrity Watch.*** The embarked unit will provide personnel to stand the air department integrity watch.

If required by the ship's CO or XO, this watch is set both underway and in port whenever there are aircraft on board and the ship is not at general quarters or flight quarters. The watch will consist of one officer and as many enlisted personnel as needed to ensure aircraft integrity. Integrity watch personnel will be indoctrinated in equipment and procedures for flight deck and hangar deck firefighting. The air officer is responsible for the integrity watch.

c. ***Army Aviation Representative to Primary Flight Control.*** The embarked unit should provide personnel as advisors to PriFly control during flight operations. The unit representative must be fully qualified in at least one type of embarked aircraft and be familiar with all unit policies. The representative also should be familiar with the day's flight schedule or mission and act as the communications link between PriFly control and the embarked unit. Selected representative should be afforded training with PriFly control, and that training should be completed before embarkation. The unit representative will be in PriFly during day Case I VFR operations. During night or Case III or IMC, the representative will be in AOCC/HDC.

Section II. Personnel Responsibilities

2-4. FLIGHT QUARTERS STATIONS

a. When directed, flight quarters stations should be manned as prescribed in the ship's watch quarter and station bill. Unit personnel will man aircraft as appropriate. Some iterations may not require that all flight quarters stations be manned. On such occasions, specific instructions are issued at the time flight quarters are set.

b. All personnel assigned working stations on the flight or hangar decks, aviation fuels, and ordnance spaces will wear flight deck safety shoes or flight boots, if available. Personnel assigned flight quarters stations on or above the hangar will wear jerseys as prescribed in Appendix F. Flight deck personnel will wear the HPG-9A cranial impact helmet or equivalent. In addition, all personnel whose duties require them to work on the flight deck will wear goggles, sound attenuators, flotation gear, a dye marker, and an adequately secured whistle and survival light. All personnel working on the hangar deck whose duties require them to work on deck-edge elevators will wear flotation gear, a dye marker, and an adequately secured whistle and survival light.

NOTE: During flight quarters, individuals wearing improper clothing will not be permitted on the flight deck without the express consent of the air officer.

c. During night flight operations, LSE or directors will use signal wands. All other personnel will use flashlights. White flashlights will not be used under amber or red flight deck-lighting conditions.

d. For planning, AOCC/HDC should be manned and the following checklist completed one and one-half hours before scheduled flight operations (commensurate with the EMCON plan in effect).

(1) Check all communications equipment (internal and external) for proper frequencies. Check CCA radar equipment, gyro repeaters, wind speed/direction indicators, and NAVAIDs for proper operation. Align the clocks. Report all discrepancies and advise the operations, tactical air, air, landing force air, and combat cargo officers if equipment failures will affect air operations. Ensure that the ready rooms are manned and ready.

(2) Establish radio communications with shore activities as applicable.

(3) Obtain a weather report for the operating area and shore stations within aircraft divert range. Advise meteorology of any special requirements for weather information during the day.

(4) Update the aircraft status board. Advise the operations, tactical air, landing force air, and combat cargo officers if aircraft availability will seriously limit air operations.

(5) Obtain PIM and check its relation to flight advisory areas and other control areas. AOCC/HDC will continuously monitored PIM.

(6) Check message traffic for information that might affect the day's operations.

(7) Check the air plan for changes; notify stations concerned.

(8) Check all status boards for completeness and accuracy.

(9) Ensure that the embarked aviation unit flight schedules have been received.

(10) Compile mission information to brief flight crews.

e. When flight quarters are sounded, the air officer ensures that the prescribed procedures are followed for inspecting and preparing for operation of optical landing aids, elevators, aviation fuel system, and crash- and fire-fighting equipment. All discrepancies will be reported to the bridge. Only the ship's commander will decide whether to conduct flight operations when discrepancies are found in any equipment. Before flight operations, the air officer will ensure that an FOD walk-down is conducted communications equipment is tested, and the required stations are manned properly.

2-5. LANDING SIGNAL ENLISTED

Under the supervision of the air officer, the LSE visually signals the helicopter pilot, helping him make a safe takeoff and/or approach and landing on the ship. The LSE directs the pilot to the desired deck spot. He ensures general safety conditions of the flight deck area, including control of the flight deck crew. He ensures that, on signal, the helicopters are started safely, engaged, launched, recovered, and shut down. The LSE also ensures that all tie-downs are removed before liftoff and properly secured after landing. Except for wave-off and hold, which are mandatory, the LSE's signals are only advisory in nature.

Section III. Aircraft Handling

2-6. FUNDAMENTALS

a. Deck space is limited and aircraft must be moved around the ship constantly for quick launching or removal to the hangar deck. Therefore, one set of ground-handling wheels must be available for each skid-equipped aircraft. Wheeled aircraft must have serviceable wheel chocks for parking on deck.

b. The current tie-down chains in the Army inventory are not compatible with the tie-down points on the decks of Navy ships. The hook on the chains is too small for the tie-down point. Although there are ways to secure the aircraft using a combination of these chains, units must ask for the proper chains during the presail conference.

NOTE: NWP 3-04.1 contains an incorrect picture of an AH-64 tie-down configuration. The mooring lugs used for C5A movements may be used on AH-64 aircraft during shipboard deployments. These lugs may remain in place during flight operations.

d. Consideration should be given to tie-down procedures during rough seas and high winds. An aircraft carrier can cruise in excess of 30 knots. When coupled with a head wind, exceeding aircraft tie-down criteria is very easy. Aircraft operator's manuals do not state maximum wind speeds for mooring. Therefore, units should consider obtaining, possibly through local manufacture, more substantial tie-downs for main rotor blades in particular. Current tie-down configurations may not be sufficient for continuous travel at sea.

e. While on board larger ships (CV/CVN or LHA/LPH/LHD), Navy personnel will maneuver (spot) all aircraft. Unit maintenance personnel and safety officers must brief the Navy handlers on what is required to move helicopters safely. During movement, Army crew chiefs will ride the brakes on wheel-landing-equipped aircraft.

2-7. HELICOPTER RECOVERY TIE-DOWN PROCEDURES

a. Given the signal from the LSE and with concurrence from the aircraft commander, chocks and tie-downs are applied after landing. They will remain attached until the aircraft is ready to take off. During short on-deck times, such as when troops or supplies are rapidly loaded, only the chocks may be applied. Tie-downs will be installed according to the individual aircraft operator's manuals.

WARNING

Winds and deck motion must be kept within the operating limits of helicopters with turning rotor blades. If the helicopter's rotor blades are turning, the, pilots will be informed before the ship starts a turn.

b. ***Personnel Debarkation.*** Pilots of helicopters with ramps will not lower the ramps to discharge passengers until the LSE gives the signal. For troop off-load, the LSE will not signal for the ramp until CCO troop handlers are present and recoveries or launches are complete on adjacent spots. As directed by the CCO, CCO handlers escort the troops from the flight deck to the troop shelters. Flight deck, flight crew, or CCO personnel will escort passengers to safe areas.

c. ***Rotor Disengagement and Engine Shutdown.***

(1) Before disengagement and/or engine shutdown, the LSE ensures that the signal to disengage is received from the flight deck officer who in turn receives the signal from the air officer. The LSE ensures that wheels are chocked, personnel are clear of rotor blades, and tie-downs are installed properly.

NOTE: The landing gear, external auxiliary fuel tank, and ordnance safety pins will be inserted before the rotor blade is disengaged and/or the engine shut down.

(2) The pilot should not disengage the rotor blade while the ship is in a turn unless authorized by the ship's commanding officer or his designated representative. The aircraft commander must be informed of wind parameters and the ship's heel before the turn starts.

WARNING

Reported winds as displayed in PriFly may vary greatly from winds blowing over the deck. Exercise extreme care when engaging or disengaging rotor blades if other aircraft are being launched or recovered. Do not attempt to engage the rotor unless the tie-down configuration is as shown in the aircraft operator's manual. Failure to comply with this requirement may induce ground resonance.

Section IV. The Air Plan

2-8. SCOPE

a. A ship's air plan is a complete daily schedule of flight operations performed on board the ship. Disseminated by the ship's operations department, the air plan becomes an order of the ship's commander.

b. Normally, Army aircraft are not under OPCON of the ship's commander. Therefore, the sequence for processing air requests and scheduling Army aircraft flights requires close coordination and cooperation between the aviation unit's scheduling authority, the CTF, and the ship's air operations officer. CTF fragmentary orders and/or unit requirements are reviewed first by the aviation unit. These requirements and plans to support task force missions are coordinated with the ship's air operations officer.

c. The air operations officer balances flight requirements against ship capabilities and the requirement to formulate the air plan. The aviation unit flight schedule is the coordinated end-product of the air plan. Distribution of the air plan and flight schedule is made according to ship requirements. All changes to the air plan and any changes to the assigned aviation unit's flight schedules that will affect the ship's air plan must be approved by the ship's air operations officer.

d. If mutually approved by the aviation unit commander and the ship's commander, the air plan may be expanded to include normal flight scheduling information provided by the aviation unit. This eliminates the requirement to publish a daily flight schedule. When this scheduling method is used, the aviation unit commander or OIC maintains authority and responsibility for scheduling assigned aircraft and crews.

NOTE: Any last-minute changes in aircraft assignment will be relayed immediately to AOCC/HDC and PriFly immediately.

2-9. CONTENTS

a. As a minimum, the air plan contains the following information:

- Event number.
- Launch time.
- Recovery time.
- Number and model of aircraft.
- Mission.
- Fuel load required.
- Call sign.
- Controlling agency.
- Circuit designator.
- Date.
- Sunrise, sunset, moonrise, moonset, and moon phase.
- Aircraft armament or ordnance loading.
- Emergency final bearing.
- Emergency marshals.

b. Additional notes may include the following data, if appropriate:

- The ready deck schedule.
- Aircraft readiness conditions prescribed by the officer in tactical command.

- Flight identification procedures in effect.
- Readiness condition of standby aircraft.
- EMCON and HERO conditions.
- Any other information required, including restrictions or hazards to flight.

c. AOCC/HDC prepares a mission brief/card for each helicopter performing a logistics mission. While not a routine mission for Army aircraft, pilots assigned the logistics mission can expect the mission brief/card to contain, at a minimum, the following information:

- Order of ships to be visited.
- Ship names, hull numbers, call signs, NAVAIDs.
- Expected bearing/distance to each ship.
- Pertinent radio frequencies.
- Number of passengers to be delivered and/or picked up and pickup and delivery points.
- Weight and description of cargo to be delivered and/or picked up.
- Certification/waiver status of ships to be visited.

2-10. MAINTENANCE TEST FLIGHTS

a. After receiving a request, the aviation unit operations officer schedules maintenance test flights through the ship's air operations officer. When feasible, these flights may be scheduled as part of routine multiple aircraft launches. When operations allow, a dedicated spot should be available to launch nonscheduled test flights.

NOTE: Test flights are prohibited during night or IMC (less than 1000/3).

b. Auxiliary power plant starts, rotor blade folding or unfolding, engine starts, and aircraft movements must be coordinated between aviation unit personnel and the ship's air department.

2-11. FLIGHT PLAN

a. Written authorization, either as an air plan, daily flight schedule, or a local flight clearance, is a prerequisite for all flights. Unscheduled flights will be kept to a minimum. The requirements for filing flight plans and advisories vary with each operating area and are contained in the Foreign Clearance Guide, flight planning documents, and fleet operating directives. Whenever possible, maintenance test flights should be scheduled on the air plan.

b. As a rule, flights originating on board the ship and terminating at a shore station, proceeding over land, or penetrating an ADIZ require the filing of a written flight plan with the ship by the pilot in command or flight leader. When firm information concerning departure and arrival times is available, the ship sends a message as soon as possible and before the ETA of the aircraft. Whenever possible, voice communication is established with the destination airfield on administrative aviation frequencies (US Air Force HF/SSB airways and command control stations, and USN/USMC Rasberry nets).

c. DD Form 175, (Military Flight Plan), ICAO, or DOD international flight plan will be filed according to the appropriate FLIP documents.

d. A departure message (IMMEDIATE precedence) is sent from the ship. The message will include the type of aircraft, aircraft bureau number, and actual time of departure. This procedure applies specifically to flights of such distance that radio communication between the ship and the aircraft will be lost before communications are established with the shore station. The time is annotated on the ship from which the flight originated. The original copy of the flight plan will be retained for three months. When the flight is completed, the PC or AMC will close his flight plan by sending an IMMEDIATE message to the ship.

e. Flight advisories will be filed for flights within ADIZ boundaries for all aircraft that will land back on board the ship and are not covered by a flight plan. Aviation units will prepare the necessary flight plans (DD Form 175 or ICAO) and file them with AOCC/HDC as far ahead of scheduled launch times as possible. AOCC/HDC files the flight plan or advisory with the appropriate agency through the available facilities.

f. While embarked, aviation units will continue to perform risk assessments according to AR 95-1.

2-12. AQUEOUS FILM-FORMING FOAM SYSTEM AND MOBILE FIREFIGHTING EQUIPMENT

The guidelines for manning and using the AFFF system are in NAVAIR 00-80R-14. Army aviation personnel are encouraged to participate in all flight and/or hangar deck fire and crash drills. These drills provide invaluable training on the AFFF system and on mobile fire-fighting equipment.

CHAPTER 3

SHIPBOARD AIR TRAFFIC CONTROL

This chapter provides information on shipboard air traffic control procedures and capabilities.

3-1. RESPONSIBILITIES

a. *Operations Officer.* The ship's operations officer is responsible for the control of airborne aircraft unless control is assigned to other authority. Control refers to all airborne operations not incidental to the actual launch or recovery of aircraft.

b. *Air Operations.* The air operations officer is responsible to the operations officer for coordinating matters pertaining to flight operations. He also ensures that the AOCC/HDC functions properly.

c. *Air Officer (Air Boss).* The air officer is responsible for visual control of all aircraft operating in the control zone. Under Case I and II conditions, this responsibility may be extended beyond the control zone to include all aircraft that have been switched to the air officer's control frequency. For special operations such as bombing a sled or conducting air demonstrations, the air officer may exercise control outside the control zone. Additionally, he is the control zone clearing authority. Agencies wanting to operate aircraft within the control zone will obtain the air officer's approval before entry. This clearance will include--

- Operating instructions for avoiding other traffic (as required).
- Information concerning hazardous conditions.
- Altitude and distance limitations at which aircraft may be operated.

d. *Combat Information Center Officer.* The CIC officer is responsible for mission control of aircraft that he is assigned. This includes providing separation from other traffic operating near the ship and ensuring that mission controllers know the basic procedures for air traffic control. He also will ensure that controllers know their responsibilities for issuing traffic advisories to aircraft operating in visual conditions and for the safe separation of aircraft operating in instrument conditions. Upon request, the CIC officer provides information concerning areas of special operations such as air-to-surface weapon drops and air-to-air missile shoots.

e. *Tactical Air Officer.* The tactical air officer controls and coordinates airborne tactical aircraft and helicopter operations with supporting arms and other air operations through the TACC (afloat).

3-2. AIRCRAFT CONTROL CRITERIA

Weather in the control zone is the most prominent factor that affects the degree of control. Unless higher authority states otherwise, the air operations officer determines the type of control used during departure and recovery.

a. *Close Proximity Operations.* Amphibious task force operations often require close proximity flight operations by two or more aviation and/or amphibious aviation assault ships. When this occurs, CVs, LPHs, LHAs, and LHDs should be assigned operating areas large enough to preclude mutual interference. At times, operational constraints may require that aviation and/or amphibious aviation ships operate within 10 nm of one another, which creates an overlap of control zones. To ensure operational safety and efficiency when these operations are anticipated, the OTC distributes special instructions (spins) that describe the limits of each ship's airspace control. These instructions also will describe the procedures used for VMC operations between contiguous control zones.

b. *Planning.* Detailed planning should be conducted to prescribe the responsibilities and procedures to be used during anticipated close proximity operations. Planning considerations should include, but are not limited to--

- Meteorological conditions (IMC or VMC).
- The type and number of aircraft (characteristics that affect control requirements).
- The type, number, and disposition of ships.
- The type of operations planned (EMCON, well-deck operations, VERTREP, refueling, and so on).
- Communications (equipment frequency availability and so on).

c. *Operations.* During concurrent flight operations by two or more LPHs/LHAs/LHDs or concurrent operations between an LPH/LHA/LHD and other aviation-capable ships (fixed-wing or rotary-wing), each ship will remain in its assigned operating area to reduce air traffic coordination problems. To avoid interference, AOCC/HOC will monitor and coordinate flight patterns. Before any aircraft operations are conducted between contiguous control zones and/or within 10 nm of the LPH/LHD, an exchange of air plans must be included in the prelaunch procedures. These procedures also must include a notification by air-capable ships and an acknowledgment by the LPH/LHA/LHD.

NOTE: Unscheduled launches or recoveries due to emergency or operational necessity are permissible. They must, however, be coordinated with the OIC as soon as possible because of the danger involved in contiguous flight operations.

d. *Electronic Control.* The five types of electronic control are close, advisory, monitor, nonradar, and electronic emission .

(1) *Close Control.* This control is used--

- During helicopter operations when the ceiling is 500 feet or less.
- During helicopter operations when the forward flight visibility is 1 mile or less.
- During all flight operations between one-half hour after sunset and one-half hour before sunrise (except as modified by the OTC or ship's commanding officer).

NOTE: Night time helicopter touch-and-go operations are excluded from close control if a visible horizon exists.

- During a mandatory letdown in thunderstorm areas.
- In other situations when supervisory personnel anticipate weather phenomena that might cause

difficulty for pilots.

(2) ***Advisory Control.*** This control is used when traffic density in an operating area requires a higher degree of control for safety of flight than required under VFR. Advisory control normally is limited to MG and is recommended for all operations in or adjacent to oceanic control areas or routes.

(3) ***Monitor Control.*** This control is used only when aircraft are operating outside controlled airspace and the pilot can safely assume responsibility for separation from other traffic.

(4) ***Nonradar Control.*** This control is used when the ship's radar is inoperative or so degraded that it cannot provide radar separation of air traffic under conditions that normally require close control. The decision to attempt to control aircraft at night or during IFR conditions will be made only after considering the--

- Actual meteorological conditions.
- Degree of radar degradation.
- Expected duration of radar degradation.
- Fuel states/fuel available for delays.
- Divert field suitability/availability.
- Operational requirement.
- Departure or recovery in progress at the time a nonradar environment develops.
- Availability of other surface or airborne platforms to provide radar traffic separation and approach information.

e. ***Electronic Emission Control.*** The ship's operations officer is responsible for EMCON and setting EMCON conditions. Special procedures may be necessary to perform the some operations during various conditions. These operations are--

- Aircraft handling.
- Launch.
- Departure.
- Mission.
- Arrival.
- Recovery.
- Maintenance.

Before conducting operations under EMCON conditions, detailed briefings must be conducted that cover responsibilities and procedures. All flight crew members, controllers, and aircraft handling personnel will attend these briefings and familiarize themselves with their procedural responsibilities. Overhead messages will include applicable EMCON instructions.

3-3. CONTROL ZONE OR CONTROL AREA LIMITATIONS

WARNING

Operating procedures in this publication that relate to the ship's control zones may not be recognized or honored by other than USN/USMC aircraft that operate from ships. Civil aircraft or aircraft from other services may enter or transit the control zone without clearance, radio contact, or regard for the procedures in this manual. They may adhere only to the basic requirements or FAR 91 (no closer than 500 feet to any vessel; less for helicopters in uncontrolled airspace). Others may not be aware of the ship's presence or its conduct of flight operations. Utmost vigilance is required in areas near airways, airfields, controlled airspace, or special-use airspace.

a. The control zone will not be effective in any portion of the area that extends into, under, or abuts controlled airspace airfields. The upper limit of the control zone must not penetrate the FCA, the floor of a TCA, or other controlled airspace. Likewise, the lateral extent is not effective in any portion that extends into or abuts controlled airspace as defined in the applicable FAA/ICAO aeronautical publications.

b. The control zone is not effective in an area that lies within special-use airspace (restricted area, MOA, and so forth) without authorization of the designated controlling agency.

c. The outer limit of the control zone will not be established closer than 10 nm to any airway, controlled airspace, or special-use airspace unless approved by cognizant authority (controlling activity, scheduling activity, or FAA facility).

d. The need may arise to activate a control zone in fleet operating areas in uncontrolled airspace. To do this, the ship's commander must coordinate with and get approval from the applicable FACSFAC, operations coordinator, numbered fleet commander, or FAA facility. He also must follow the same procedures to activate a control zone in underlying airways or controlled airspace or adjacent to special-use airspace.

3-4. AIRCRAFT SEPARATION CRITERIA

The following separation standards are used for aircraft under close control. These restrictions do not apply to tactical maneuvers such as air intercept rendezvous and close ASW action.

a. ***Lateral Separation.*** The following separation standards apply to aircraft controlled by designated air search radars:

(1) Aircraft operating 50 miles or more from the monitoring antenna must be separated by a minimum of 5 miles.

(2) Aircraft operating within 50 miles of the monitoring antenna and are not within 10 miles on a designated approach must be separated by a minimum of 3 miles.

(3) Aircraft on a designated approach and inside 10 miles must be separated by a minimum of 2 miles.

(4) Aircraft established on final within 5 miles must be separated by a minimum of 1 1/2 miles.

Aircraft given positive separation through nonradar control using a published approach/departure will be separated by two minutes (5 miles separation when using DME).

b. ***Vertical Separation.***

(1) Jet and turboprop aircraft operating at altitudes up to and including FL 290 will be separated by 1,000 feet vertically. Aircraft operating at altitudes above FL 290 will be separated by 2,000 feet vertically.

(2) Helicopters will be separated by 500 feet vertically. Helicopters will be separated from fixed-wing aircraft by 1,000 feet vertically.

3-5. COMMUNICATIONS CONTROL

a. Unless otherwise directed, all aircraft will be under positive communications control while operating at sea. Pilots will not shift frequencies without notifying and/or obtaining clearance from the controlling agency. Communications procedures during ZIP-LIP/EMCON conditions will be specified during the preflight briefing. The following paragraphs outline how control of radio circuits is exercised.

(1) ***AOCC/HDC.***

- Primary control of assigned ship-to-shore ATC and intratype administrative frequencies.
- Primary control of assigned GCA frequencies.
- Primary control of helicopter direction (tactical) frequencies.
- Secondary control of aircraft guard frequencies.
- Secondary control of land/launch frequencies.
- Secondary control of tactical air frequencies.

(2) CIC/TACC.

- Primary control of all air tactical frequencies not otherwise assigned.
- Primary control of aircraft guard frequencies.
- Secondary control of ship-to-shore ATC and intratype administrative frequencies.

(3) ***PriFly.***

- Primary control of land/launch frequencies.
- Secondary control of aircraft guard frequencies.
- Secondary control of departure control and final approach frequencies.
- Secondary control of assigned ship-to-shore ATC and intratype administrative frequencies (where installed equipment permits).

b. Strict radio discipline is mandatory. Voice procedures must be concise and should not vary from standard air control phraseology given in Navy regulations (ACP 165 and current OPNAVINST 3721.1-series). The unit or ship call sign and the aircraft side number or alphanumeric call signs will be used exclusively after initial contact. Radio circuits that are used to control air traffic are recorded continuously during hours of operation.

c. Communications security is best accomplished by strict adherence to the principles of radio discipline. In addition, equipment in naval aircraft and ships offer a significant COMSEC capability that should be used to the greatest extent practicable. All units with COMSEC capability should develop tactical doctrine designed to deny SIGINT forces access to vital intelligence. Detailed functional descriptions of COMSEC equipment are found in pertinent classified documents. All personnel who have access to radio equipment must be briefed that certain restrictions are placed on all radio transmissions to prevent disclosure of EEFI.

3-6. EMERGENCY CONTROL PROCEDURES

From a control standpoint, emergencies fall into five categories. These categories are communications failure, navigational aids failure, aircraft systems failure, crew member injury or illness, and ship system casualty. The resolution of an emergency involves a command decision based on the type of emergency and the weather conditions in the recovery area. AOCC/HDC must collect every detail that might help evaluate an emergency and keep the command and other interested agencies properly informed. The following paragraphs discuss the basic procedures to follow when communications and navigation equipment fail. Emergencies that occur when navigation aids and/or communications are operational should be handled according to existing circumstances. Emergency procedures for aircraft system failures are covered in the appropriate NATOPS flight manual.

a. *Initial Control Responsibility.* The agency that has control of the aircraft when the emergency occurs has initial control responsibility. Aircraft in distress should not change radio frequencies if satisfactory radio contact is established. Neither should controllers require that aircraft in distress change frequencies.

b. *Basic Emergency Control Procedures.* When pilots experience communications and/or navigation equipment failure, they should follow procedures consistent with Army and FAA regulations. Controlling agencies will be alert for conditions that indicate communications or navigation failures. As appropriate, these agencies should--

- Attempt to establish communications with the aircraft.
- Attempt to establish control of the aircraft.
- Vector the aircraft.

If the controlling agency cannot communicate with the aircraft,--

- Identify the aircraft on radar and maintain a track.
- Vector available aircraft to join, if practical.
- Clear all other aircraft from the track of the distressed aircraft.
- Broadcast instructions and essential information in the blind.

c. *Crew Member Injury or Illness.* If a crew member is injured or becomes ill, state the nature of the injury or illness, request assistance, and state your intentions to the controlling agency. Under these conditions, aircraft normally are handled as an emergency and vectored for immediate recovery. When divert or recovery is not possible, ditching procedures are performed according to the aircraft operator's manual..

d. *Ship System Casualty.* A ship system casualty can result in complete shipboard communications equipment and navigational aid failure. Certain casualties may result in the inability to maneuver the ship to the BRC and provide optimum winds. Pilots must be familiar with the conditions that indicate a ship system casualty and perform the following actions, as appropriate:

- Attempt to establish communications and coordination with other aircraft.
- Enter Charlie pattern and obey visual signals.
- Execute divert procedures.
- Execute ejection or ditching procedures according to the aircraft NATOPS flight manual.

3-7. TRANSIENT AIRCRAFT

The controlling agency will advise the aircraft of the BRC and/or all course changes. Transient aircraft approaching the ship for landing will contact AOCC/HDC at least 25 miles out or when "Feet Wet."

3-8. LOST AIRCRAFT PROCEDURES

a. When the position of an aircraft is in doubt, the controller will immediately begin the procedures outlined below.

(1) Attempt to obtain radio or radar contact. Use a relay aircraft to attempt radio contact on the circuit in use and on guard frequencies. Continue to send information in the blind and search all IFF modes. Begin a communications search and monitor the guard channels (243.0 and 40.50) for emergency aircraft calls.

(2) Inform the OCE/OTC.

(3) Keep an estimate of the aircraft's fuel state.

(4) Check the weather and clear airspace for emergency marshal as required.

(5) Check to ensure that navigation aids are operable. If navigation aids are inoperable, alert the command for the possible use of other aids to lost aircraft such as search aircraft, black smoke, vertical searchlights, antiaircraft bursts, starshells, fire control tracking balloons, and an energized prebriefed sonobuoy channel.

(6) If communications or radar contact cannot be regained before expiration of the aircraft's last known fuel state, activate the command SAR plan.

b. When contact is regained, perform the following:

(1) Check the fuel state.

(2) Vector the aircraft to the ship or divert, as appropriate.

(3) Vector the aircraft for escort, if necessary.

(4) Maintain regaining contact track of the aircraft.

(5) If communications are unsatisfactory, use a relay aircraft or have the lost aircraft gain altitude, if possible.

CHAPTER 4

LAUNCHING AIRCRAFT

This chapter provides guidance for launching aircraft on missions from amphibious ships (LPH, LHA, and LHD) and aircraft carriers (CV and CVN). Operations from single-spot ships are covered in Appendix D.

Section I. General Information

4-1. OPERATIONAL PROCEDURES

During all phases of flight operations, positive communications will be maintained among the flight deck, HDC, PriFly, and bridge. This will ensure that the OOD controls the ship so that wind and deck motion remain within the prescribed envelope. During all phases of air operations, the OOD will inform PriFly and AOCC/HDC before changing the BRC and speed. The OOD also will provide the expected BRC and speed. The ship must be kept on a steady course and speed during rotor engagement or disengagement, engine start and shutdown for aircraft without rotor brakes, taxiing, and launch or recovery operations. Deck tilt, centrifugal force, or rapidly changing wind direction or velocity aerodynamically affects the controllability of the aircraft and may cause rollover. Permission must be obtained before the movement, engagement, disengagement, launch, or recovery of any aircraft. As the representative of the ship's CO, the OOD and the air officer have supervisory responsibility for safe operations.

a. ***Time Schedule***. All flight preparations will be completed in time for the pilots to conduct preflight inspections of their aircraft before the scheduled launch time. Every effort must be made to prevent delays in the launch cycle.

b. ***Flight Quarters***. Flight quarters must be set in time for all personnel to man stations and prepare for flight operations. When flight quarters are set, the following stations report to the OOD or air officer, as appropriate.

- PriFly.
- Hangar deck.
- Flight deck.
- Aviation fuels.
- AOCC/HDC.
- Rescue boat detail.
- Crash crew and firefighters.
- CIC.
- Medical crew.

• Mine countermeasures launch crew (when embarked).

c. *Primary Flight Control*. PriFly provides recovery, launch, and operational control of aircraft while they are on the ship and within the ship's control area (Figure 4-1). PriFly interfaces with AOCC/HDC to control airborne aircraft and with the CCO to integrate assault elements with helicopters on the flight deck. On-ship control of aircraft includes spotting, maintenance, fueling and defueling, arming and dearming, movement, stowage, and aircraft handling on the flight and hangar decks.

NOTE: CVs and CVNs do not have a CCO. The officer in charge of air transportation is the ATO. The ATO works for the air operations officer. He coordinates the transport of all passengers, mail, and cargo that is flown off the deck in helicopters and in fixed-wing aircraft.

d. *Communications*. PriFly has many communications terminals, both internal and external. Internal communications systems link PriFly with other ship control spaces, and internal radio systems provide communications control of personnel on the flight deck. PriFly also controls the 5 MC (flight deck announcing system). Communications equipment provides PriFly with radio and visual (Aldis lamp) links to aircraft under PriFly control.

e. *Flight Deck Lighting and Optical Landing Aids*. PriFly controls optical landing aids and flight deck lighting.

f. *Helicopter Readiness Conditions*. Flight crews assigned the alert conditions discussed in the following paragraphs will be called away early enough to allow for a normal preflight inspection, start, and run-up and completion of the takeoff check by the time stated in the air plan. After the pilot declares the helicopter ready for flight, it is placed in the appropriate alert condition.

(1) *Condition I/Alert 5*. The helicopter will be spotted for immediate launch with the rotor blades spread; starting equipment plugged in; and the LSE, starting crew man, and ordnance personnel ready for launch. When the word is passed to "stand by for launch," engines will be started without further instructions. However, rotor engagement and launch (engine start for aircraft without rotor brakes) will be positively controlled from PriFly. Aircraft should be airborne within five minutes of the order to launch.

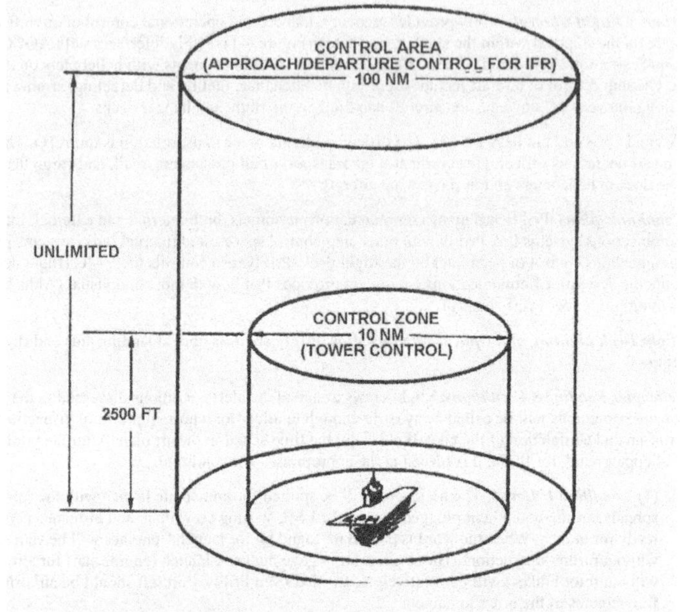

Figure 4-1. Control area and control zone dimensions

(2) ***Condition II/Alert 15.*** The same conditions apply as for Condition I, except that flight crews are not required to be in the helicopter. They will be on immediate call. Aircraft should be airborne within 15 minutes of the order to launch.

(3) ***Condition III/Alert 30.*** Main rotor blades may be folded. The helicopter need not be in position for immediate launch; however, it must be parked to allow direct access to a suitable launch spot. A tow bar (if required) will be attached to the helicopter and a specific LSE, tractor driver, handling crew, and starting crew man will be assigned to each helicopter. These personnel must be briefed thoroughly so that when the order is given to prepare to launch, the helicopter can be quickly and safely moved into position and readied for launch. Flight crews will be in their flight gear, prebriefed for the launch, and in the ready rooms or working spaces. Aircraft should be airborne within 30 minutes of the order to launch.

(4) ***Condition IV/Alert 60.*** This condition is similar to Condition III, except that minor maintenance may be performed if no restoration delay is involved. The aircrew will be designated and available. Aircraft should be airborne within 60 minutes of the order to launch.

4-2. AIR OFFICER AND AVIATION UNIT OPERATIONS DUTY OFFICER RESPONSIBILITIES

a. ***Air Officer.*** The air officer is responsible to the ship's CO for activities that support flight operations on the flight deck and hangar deck. The air officer or a qualified assistant will be in PriFly during flight quarters to control all evolutions involving aircraft. The air officer will confirm aircraft assignments with the AOCC or HDC and the unit maintenance controller or liaison officer before respotting the flight or hangar decks for launch. In addition to the ship's air plan, the air officer also will maintain an up-to-date copy of the unit flight schedule in PriFly. During Case III/night operations, both PriFly positions will be manned. One of these positions will be manned by either the air officer or assistant air officer.

b. ***Aviation Unit Operations Duty Officer.*** The ODO is responsible to the unit operations officer for coordinating and executing the flight schedule. During flight quarters, he will remain in the ready room and monitor the applicable communications circuits. He will keep the AOCC or HDC and PriFly (if necessary) notified of any changes that may affect launch or recovery operations.

Section II. Flight Deck Procedures

4-3. GENERAL PROCEDURES

a. ***Flight Deck Description.***

(1) The flight deck is marked with eight spots for the LPH (Figure 4-2). Markings also are provided for "Spot Mike," a designated landing spot for aircraft conducting airborne mine (AMGM) sled operations. The flight deck is marked with ten spots for the LHA (Figure 4-3) and nine spots for the LHD (Figure 4-4). Appendix D contains FFG, DD, and DDG deck markings.

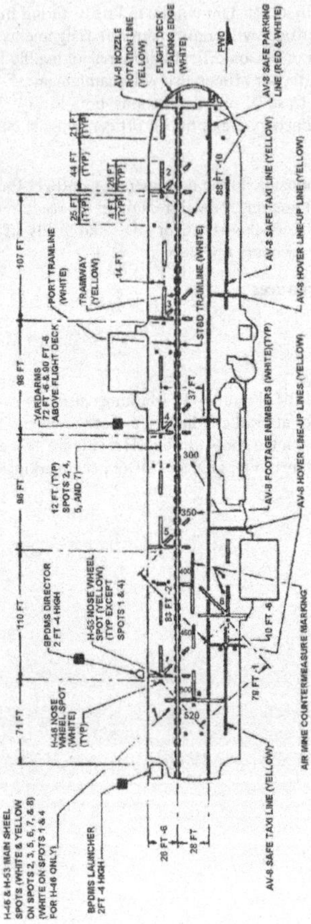

LPH HELICOPTER SPOTTING MATRIX

HELO SPOT NO.	HELICOPTER SPOTTING ARRANGEMENTS											
	NO. 1	NO. 2	NO. 3	NO. 4	NO. 5	NO. 6	NO. 7	NO. 8	NO. 9	NO. 10	NO. 11	NO. 12
1	H-46	H-46	H-46	H-46	H-46	H-46	H-46	H-46	H-46	—	—	H-46
2	H-46	H-53 A/D	H-53A/D	H-53A/D	H-46	H-46	H-53 A/D	H-53 A/D	H-46	H-53E	H-53E	H-46
3	H-46	H-53 A/C	H-46	H-46	H-53A/D	H-46	H-53 A/D	H-46	H-53E	—	—	H-53E
4	H-46	—	H-46	H-46	—	H-46	—	H-46	—	H-46	H-46	—
5	H-46	H-53 A/D	H-53A/D	H-53A/D	H-46	H-53 A/D	H-46	H-46	H-53E	—	H-53E	—
6	H-46	—	—	H-46	—	H-46	H-46	—	—	H-53E		H-53E
7	H-46	NOTE 2	NOTE 2	H-46	NOTE 2	NOTE 2	H-46	NOTE 2	NOTE 3	—	NOTE 3	—
8	—	NOTE 2	NOTE 2	—	NOTE 2	NOTE 2	—	NOTE 2	NOTE 3	—	NOTE 3	—

NOTES:

1. THE HELICOPTER SPOTTING MATRIX IDENTIFIES VARIOUS US NAVY/MARINE CORPS AIRCRAFT MIXES THAT SHOULD BE USED TO SATISFY MULTISPOT OPERATIONAL REQUIREMENTS. IT HAS BEEN DEVELOPED BASED ON AVAILABLE ROTOR CLEARANCES AND DOES NOT CONSTITUTE AUTHORITY TO OPERATE. FURTHERMORE, THE AIR OFFICER IS RESPONSIBLE FOR DETERMINING WHAT ARRANGEMENT CAN BE USED FOR GIVEN ENVIRONMENTAL CONDITIONS.

2. ARRANGEMENTS 2, 3, 5, 6, AND 8: H-53A/D/E HELICOPTERS MAY OPERATE ON EITHER SPOTS 7 OR 8.

3. ARRANGEMENTS 9 AND 11: H-53A/D HELICOPTERS MAY OPERATE ON SPOT 7, OR H-53A/D/E HELICOPTERS MAY OPERATE ON SPOT 8.

4. USN/USMC H-1 OR H-2 HELICOPTERS MAY BE SUBSTITUTED FOR ANY H-46 IN THE MATRIX. USN/USMC H-1, H-2, H-3, OR H-60 HELICOPTERS MAY BE SUBSTITUTED FOR ANY H-53A/D IN THE MATRIX. USN/USMC H-1, H-2, H-3, H-46, H-53A/D, OR H-60 HELICOPTERS MAY BE SUBSTITUTED FOR ANY H-53E IN THE MATRIX. IN ALL CASES, THE FORWARD SKID EXTREMITIES OF ALL H-1 HELICOPTERS AND THE NOSE OF ALL H-2, H-3, AND H-60 HELICOPTERS SHALL BE ON/OVER THE ATHWARDSHIP LINE-UP LINE.

5. NOMINAL ROTOR CLEARANCES FOR THE AIRCRAFT MIXES ADDRESSED IN THE MATRIX EQUAL OR EXCEED 15 FEET, WITH THE EXCEPTION OF THE FOLLOWING:

a. H-46 ON SPOT 1 TO H-53A/D ON SPOT 2: 12 FT.
b. H-46 ON SPOT 1 TO H-46 ON SPOT 2: 12 FT-4.
c. H-53A/D ON SPOT 2 TO H-46 ON SPOT 3: 14 FT-6.
d. H-46 ON SPOT 3 TO H-46 ON SPOT 4: 13 FT-8.
e. H-46 ON SPOT 4 TO H-46 ON SPOT 5: 13 FT-8.
f. H-46 ON SPOT 4 TO H-53E PM S[PT 5: 14 FT-2.
g. H-46 ON SPOT 5 TO H-46 ON SPOT 6: 13 FT-6.
h. H-53A/D ON SPOT 5 TO H-46 ON SPOT 6: 12 FT-2.
i. H-53E ON SPOT 5 TO H-53A/D ON SPOT 7: 14 FT-4.
j. H-46 ON SPOT 6 TO H-46 ON SPOT 7: 11 FT-6.
k. H-53A/D (TAIL ROTOR) ON SPOT 7 TO BPDMS: 10 FT-6.
l. H-53E (TAIL ROTOR) ON SPOT 7 TO BPDMS: 8 FT-3.

THESE CLEARANCES ARE PUBLISHED FOR INFORMATION PURPOSES ONLY. THEY MAY VARY BETWEEN SHIPS AND MAY BE REDUCED OR EXCEEDED DEPENDING UPON ACTUAL INDIVIDUAL SPOT LOCATIONS AND ISLAND APPENDAGES.

6. FOR CURRENT HELICOPTER SPOT/AV-8 MARKING ARRANGEMENT OF EACH LPH CLASS SHIP, SEE THE SHIPBOARD AVIATION FACILITIES RESUME (NAVAL AIR ENGINEERING CENTER REPORT NAEC-ENG-7576).

7. THE SYMBOL ■ DENOTES FLIGHT CAUTION ITEM.

Figure 4-2. LPH typical arrangement of helicopter spots

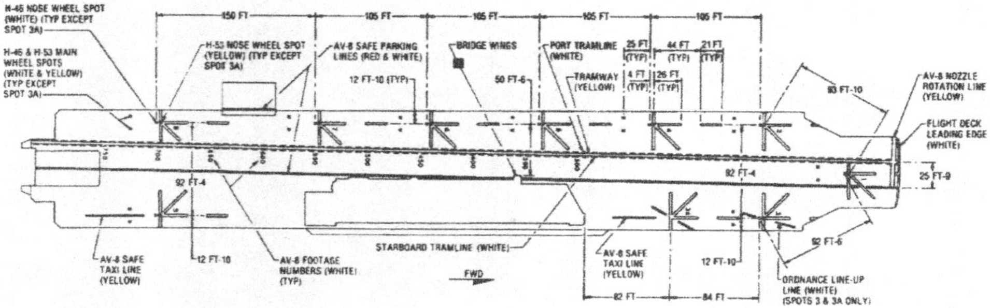

LHA HELICOPTER SPOTTING MATRIX

HELO SPOT NUMBER	HELICOPTER SPOTTING ARRANGEMENTS							
	NO. 1	NO. 2	NO. 3	NO. 4	NO. 5	NO. 6	NO. 7	NO. 8
1	H-46	H-53A/D	H-46 or H-53A/D	H-53E	H-53E	—	H-46 or H-53A/D	—
2	H-46	H-53A/D	H-46 or H-53A/D	H-46	H-46	H-53E	H-53E	—
3	H-46	H-53A/D	H-46 or H-53A/D	H-46	H-46	H-53E	H-46 or H-53A/D	—
3A	NOTE 3	NOTE 3	NOTE 3	NOTE 3	NOTE 3	NOTE 3	NOTE 3	—
4	H-46	H-53A/D	H-46 or H-53A/D	H-53E	H-46 or H-53A/D	—	—	—
5	H-46	H-53A/D	H-46 or H-53A/D	—	H-53E	H-53E	H-53E	—
6	H-46	H-53A/D	H-46 or H-53A/D	H-53E	—	—	—	H-46
7	H-46	H-53A/D	H-46 or H-53A/D	—	H-53E	H-53E	H-53E	H-53E
8	H-46	H-53A/D	H-46 or H-53A/D	H-53E	H-53E	H-53E	H-53E	—
9	H-46	H-53A/D	H-46 or H-53A/D	H-53E	H-53E	H-53E	H-53E	—

NOTES:

1. THE HELICOPTER SPOTTING MATRIX IDENTIFIES VARIOUS U.S. NAVY/MARINE CORPS AIRCRAFT MIXES THAT SHOULD BE USED TO SATISFY MULTI-SPOT OPERATIONAL REQUIREMENTS. IT HAS BEEN DEVELOPED BASED ON AVAILABLE ROTOR CLEARANCES, AND DOES NOT CONSTITUTE AUTHORITY TO OPERATE. FURTHERMORE, IT IS THE RESPONSIBILITY OF THE AIR OFFICER TO DETERMINE WHAT ARRANGEMENT CAN BE UTILIZED FOR GIVEN ENVIRONMENTAL CONDITIONS.

2. USN/USMC H-1 OR H-2 HELICOPTERS MAY BE SUBSTITUTED FOR ANY H-46 IN THE MATRIX. USN/USMC H-1, H-2, H-3, OR H-60 HELICOPTERS MAY BE SUBSTITUTED FOR ANY H-53A/D IN THE MATRIX. USN/USMC H-1, H-2, H-3, H-46, H-53A/D, OR H-60 HELICOPTERS MAY BE SUBSTITUTED FOR ANY H-53E IN THE MATRIX. IN ALL CASES, THE FORWARD SKID EXTREMITIES OF ALL H-1 HELICOPTERS AND THE NOSE OF ALL H-2, H-3, AND H-60 HELICOPTERS SHALL BE ON/OVER THE ATHWARTSHIP LINE-UP LINE.

3. SPOT 3A IS INTENDED FOR H-1 HELICOPTERS ONLY. IT SHOULD BE UTILIZED ONLY WHEN SPOT 3 IS VACANT OR OCCUPIED BY ANOTHER H-1 HELICOPTER. WHEN SPOT 3 IS OCCUPIED BY AN H-46 OR H-53 HELICOPTER, SPOT 3A SHOULD REMAIN VACANT.

4. NOMINAL ROTOR CLEARANCES FOR THE AIRCRAFT MIXES ADDRESSED IN THE MATRIX EQUAL OR EXCEED 15 FEET, WITH EXCEPTION TO THE FOLLOWING:

a. H-46 ON SPOT 1 TO H-53E ON SPOT 2: 14 FT
b. H-46 ON SPOT 1 TO H-46 ON SPOT 3: 14 FT-6
c. H-53A/D ON SPOT 1 TO H-53E ON SPOT 2: 14 FT
d. H-53A/D ON SPOT 1 TO H-53A/D ON SPOT 3: 14 FT-3
e. H-53E ON SPOT 1 TO H-46 ON SPOT 2: 14 FT
f. H-53E ON SPOT 1 TO H-46 ON SPOT 3: 13 FT-5
g. H-53E ON SPOT 2 TO H-53E ON SPOT 3: 13 FT-4
h. H-53A/D ON SPOT 2 TO H-46 ON SPOT 4: 12 FT-6
i. H-53A/D ON SPOT 4 TO H-46 ON SPOT 5: 12 FT-6
j. H-53A/D ON SPOT 4 TO H-53E ON SPOT 5: 13 FT
k. H-53E ON SPOT 5 TO BRIDGE WINGS: 13 FT-4 (11 FT WHEN HOVERING ABOVE SPOT 5)
l. H-53A/D ON SPOT 5 TO BRIDGE WINGS: 16 FT-8 (14 FT-4 WHEN HOVERING ABOVE SPOT 5)
m. H-53A/D ON SPOT 5 TO H-46 ON SPOT 6: 12 FT-6
n. H-53A/D ON SPOT 6 TO H-46 ON SPOT 7: 12 FT-6
o. H-53E ON SPOT 6 TO FLIGHT DECK CONTROL: 12 FT-10
p. H-53E ON SPOT 8 TO H-53E ON SPOT 9: 13 FT-4

THESE CLEARANCES ARE PUBLISHED FOR INFORMATION PURPOSES ONLY. THEY MAY VARY BETWEEN SHIPS, AND MAY BE REDUCED OR EXCEEDED DEPENDING UPON ACTUAL INDIVIDUAL SPOT LOCATIONS AND ISLAND APPENDAGES.

5. FOR THE CURRENT HELICOPTER SPOT/AV-8 MARKING ARRANGEMENT OF EACH LHA CLASS SHIP, SEE THE SHIPBOARD AVIATION FACILITIES RESUME (NAVAL AIR ENGINEERING CENTER REPORT NAEC-ENG-7576).

6. THE SYMBOL ■ DENOTES FLIGHT CAUTION ITEM.

Figure 4-3. LHA typical arrangement of helicopter spots

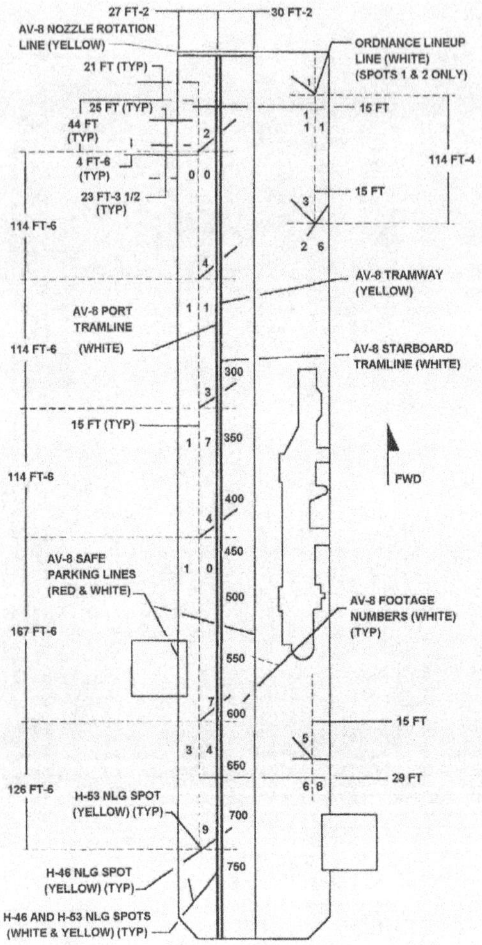

Notes:

1. This flight deck marking arrangement is typical of all LHDs. NAVAIRENGCEN Report No. NAEC-ENG-7576, entitled "Shipboard Aviation Facilities Resume" contains specifics on the arrangement of each LHD.

2. All Navy/Marine Corps helicopters can be operated from all spots during multispot operations.

Figure 4-4. LHD typical arrangement of helicopter spots

(2) The flight deck is divided into two separate landing areas. During aircraft operations, the forward area consists of spots 1 through 4 for the LPH, spots 1 through 5 for the LHA, and spots 1 through 5 for the LHD (Figure 4-4). The after area consists of spots 5 through 8 for the LPH, spots 6 through 9 for the LHA, and spots 6 through 9 for the LHD. The two landing areas are controlled separately by rotary beacon lights or flags from PriFly (Figure 4-5). A typical landing spot is shown in Figure 4-6. In addition, each spot is assigned an LSE or director who wears a helmet equipped with a transmitter/receiver unit that provides direct communications with PriFly and flight deck control.

EVOLUTION	FLAG DISPLAY	MEANING
Setting of flight quarters	HOTEL/FOXTROT flag at the dip (as appropriate)	Ship ready to conduct flight operations when wind conditions are suitable
Ready to conduct flight operations	HOTEL/FOXTROT close-up (as appropriate)	Ready to conduct or conducting flight operations
A delay or interruption of the evolution	HOTEL/FOXTROT at the dip (as appropriate)	Flight operations temporarily delayed
No flight operations being conducted	HOTEL/FOXTROT flag hauled down (as appropriate)	No flight operations being conducted

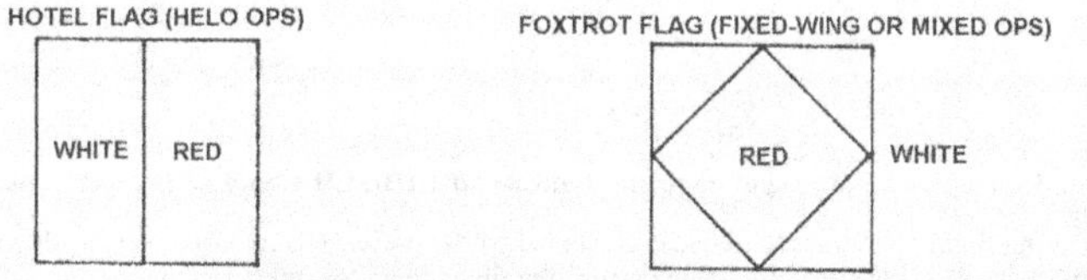

NOTE: HOTEL/FOXTROT flag is displayed just forward and above Pri-Fly

Figure 4-5. Flag hoist signals

b. ***General Flight Deck Safety.***

(1) The ship's CO is responsible for the safety of embarked aircraft and personnel. The CO or OIC of the aviation unit and the pilots of individual aircraft are responsible for the safety of assigned aircraft and personnel.

(2) The CO or OIC of the aviation unit and ship's company personnel will evaluate the hazards involved in shipboard flight operations and develop the appropriate safety measures. All personnel will be trained in safe operating procedures before flight operations begin.

WARNING

High winds, high noise levels, fire hazards, flying objects, turning rotors, taxiing aircraft, intake suction, and jet blast make safety consciousness imperative.

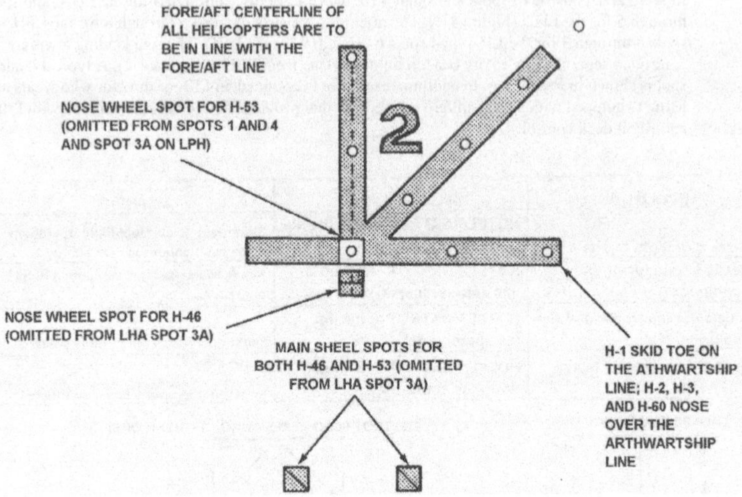

Figure 4-6. Helicopter landing spot diagram (typical) for LHD, LHA, and LPH class ships

(3) During flight operations, only those personnel whose presence is required will be allowed on the flight deck. All others will remain clear of the flight deck, catwalks, and gun tub areas. Personnel may view flight operations only from an area designated by the CO.

(4) Personnel engaged in flight operations will wear the appropriate safety helmets, sound suppressers, safety goggles, flight deck safety shoes, long sleeve shirts or jerseys, and life vests. Flight quarters clothing will conform to the colors and symbols prescribed in Appendix F. Questions about the availability of special clothing should be addressed to the Navy representative at the presail conference.

(5) While flight operations are being conducted, personnel on exposed decks and catwalks will remove all loose items of clothing and equipment including their hats (except for approved, properly fastened safety helmets).

(6) Personnel on the flight deck will be trained to take cover immediately on command of the flight deck officer, air officer, or launch officer.

(7) Personnel working near an aircraft will observe the aircraft carefully for signs of malfunction, such as smoke, oil, and hydraulic leaks. Malfunctions will be reported immediately to the flight deck officer or air officer.

(8) Crew members, passengers, and troops returning from flight will quickly clear the flight deck and the island structure exposed to flight operations. Qualified personnel will escort all passengers and troops to and from the aircraft.

WARNING

During deck and flight operations, smoking is not permitted on the flight deck, hangar deck, catwalks, elevators, or weather decks. Matches and cigarette lighters will not be used in compartments where fuel fumes may be present. The ship's CO may designate certain smoking areas above board.

(9) Dawn, dusk, and night operations increase the hazards to personnel on the flight decks. Greater vigilance is required during these periods.

(10) When aircraft are serviced, especially at night, extreme care must be taken to prevent overfilling fuel tanks and spilling oil or hydraulic fluid. Spilled oil, grease, hydraulic fluid, and fuel will be removed from the flight deck immediately.

(11) Care will be taken when approaching elevator openings, particularly on the windward side. No one will try to get on or off an elevator when the elevator operator raises the elevator stanchions. Personnel will not lean on the elevator guard rails at any time. Guard rails are in place across cargo elevator openings at all times when the main hatches are open and cargo is not being moved into or out of the elevator.

(12) Crash crew and/or organized fire parties are responsible for responding to aircraft crashes and fires. Unless specifically asked to help combat a fire or clear the deck, personnel will stay clear of the area where the fire or crash occurred. In case of fire, designated Army personnel will help handle hoses and personnel casualties. Fighting a flight deck or hangar deck fire is an all-hands evolution. Maximum participation is essential during flight deck or hangar deck drills.

(13) Care will be taken when aircraft are being spotted or parked near energized antennas. Enough voltage may be induced in the airframe to create a safety hazard.

(14) No aircraft will be spotted so that it extends over a gun tub or missile launcher.

CAUTION

No aircraft will be closer than 30 feet to any gun mount during live-fire exercises. Damage to the aircraft skin, windows, and ramps may result from overpressurization.

c. *Foreign Object Damage Hazard.* All deck areas and particularly the flight deck will be inspected before flight operations. They also will be monitored throughout flight operations to ensure that they are clear of foreign objects, such as rags, pieces of paper, line, caps, nuts, and bolts. These objects can be caught by air currents and cause damage to aircraft or injury to personnel.

WARNING

Dumping trash during flight operations creates a serious FOD hazard. The dumping of trash will stop before flight operations and not resume until flight operations are completed.

d. *Helicopter Safety Precautions.* Listed below are some helicopter safety precautions that all personnel will observe during shipboard operations.

- Personnel will not approach or depart a helicopter while the rotors are being engaged or disengaged.

- Helicopters should not routinely be deck-taxied on the flight deck.
- Helicopters will not be towed or pushed while the rotors are engaged.
- Pilots will not fly helicopters over other aircraft on launch.
- Only spots that afford visual reference to the deck will be used for night helicopter launches.
- Personnel that must be in the area of operating helicopters will exercise extreme caution and observe the signals and directions of the LSE or combat cargo representative, as appropriate.
- Dual-engine helicopters will not be intentionally hovered single engine over a deck spot. If topping checks cannot be performed on the deck, they must be performed in flight at an appropriate altitude.

e. ***V/STOL Aircraft Safety Precautions.***

(1) V/STOL aircraft (AV-8 Harrier) engines are extremely susceptible to FOD. The engine can ingest debris, which may cause the loss of an engine and possibly the loss of the aircraft.

(2) Personnel can be blinded by foreign objects propelled by aircraft jet blast.

(3) Exhaust gases from V/STOL aircraft have tremendous speed and impact force. Special precautions must be taken to remove or thoroughly secure all loose items. These items may include missile or gun director covers, deck drain covers, life raft covers, or padeye covers that are near the landing area or approach path.

WARNING

1. During V/STOL operations, a large amount of high-velocity gas is emitted downward from the exhaust nozzles. This downwash strikes the flight deck and flows horizontally above the deck, endangering the flight deck crew. Movement in this high-velocity blanket is impeded very little and is similar to walking in a swift stream of knee-deep water. However, if a flight deck crew man should fall, he may be blown overboard.

2. When the AV-8 is below 10 feet, the jet efflux produced during vertical operations will exceed 200 degrees F (93 degrees C) at a distance of 25 feet from the center of the landing spot. Flight deck personnel will remain clear of this area during takeoffs and landings. Flight deck personnel also will remain clear of the wing tips, nose, and tail area because of the jet blast danger from the reaction control ducts. (There is no blast from the reaction control ducts with the nozzles aft.) The reaction control ducts also present a hazard when the engine is off because they have sharp edges and retain heat after the aircraft is shut down.

3. The blast patterns of the AV-8 create a hazard not only to personnel and equipment on the deck but also to the aircraft. All FOD must be cleared from the flight deck and from padeyes and catwalks before AV-8 operations. Equipment, such as warning signs, hoses, and hatches, must be fastened down securely.

f. ***Burns.*** Burns from the exhausts and ducts of the AV-8 aircraft are a hazard. The deck and other objects around the aircraft become extremely hot after only a brief exposure to the exhaust gases. Flight deck personnel will be briefed thoroughly on these hazards and how to avoid them.

4-4. PREFLIGHT INSPECTIONS

a. While aircraft are still packed and waiting for deck spotting, as much of the preflight inspection will be completed as possible. All preflights normally will be completed 30 minutes (or as required) before launch time. Pilots will be strapped in the aircraft with as much of the prestart checklist completed as possible.

WARNING

Maintenance on or preflight of any portion of an aircraft that extends over the edge of the deck of the ship is prohibited.

b. Performing the preflight on areas of the aircraft that are inaccessible, such as areas over the edge of the deck, will be done after the aircraft is spotted. All aircrew and maintenance personnel will wear a safety cranial or flight helmet when climbing on a helicopter or V/STOL aircraft. Flotation gear will be worn whenever the aircraft is on the flight deck.

4-5. PRELAUNCH PROCEDURES

a. ***Launch Responsibilities.*** The OOD will set flight quarters in time for all personnel to man their stations and complete preparations for flight.

NOTE: Starting, engagement, launch, and recovery wind envelopes will be available for the OOD and air officer during flight operations.

(1) Communications circuits will be manned as appropriate.

(2) The OOD will ensure that the rescue boat is fully prepared and the boat crew detailed and available for launch. On aviation ships, such as the CV, CVN, LHA, LHD, and LPH, the boat crew usually is replaced by an airborne SAR helicopter or by one that is in an alert status.

(3) The air officer will ensure that obstructions, such as weapons, antennas, cranes, flagstaffs, and lifelines, are lowered, trained clear, or unrigged.

(4) Before the engines are started, the aircraft handling officer will ensure that a complete FOD walk-down of the flight deck and adjacent topside area is conducted.

(5) The air officer will require that all flight deck personnel use the appropriate flight deck clothing and equipment.

(6) The air officer will clear the flight deck of all unnecessary personnel.

(7) The flight deck officer will ensure that mobile crash and fire-fighting equipment is manned and ready.

(8) The OOD will display Hotel/Foxtrot at the dip and a red deck signal to PriFly as shown in Figure 4-5.

(9) Within established wind limitations, the OOD will maneuver the ship to obtain favorable wind conditions. Whenever possible, optimum winds will be provided. When environmental conditions or the ship's motion dictates, these wind limitations will be reduced to provide safe engine start, engagement or disengagement, launch, and recovery winds. The unit or detachment CO will ensure that limitations more restrictive than those established by NATOPS are discussed and agreed upon with the CO of the ship.

b. *Launch Preparation.*

(1) When an aircraft is being spotted for launch, the LSE, director, crew chief, or PC will ensure that the parking brakes are set, wheels are chocked, tail or nose gear is locked (as applicable), and safe rotor or wing clearance exists. Instructions for tie-downs will be according to the operator's manual. When specific guidance is not available, chains attached to fuselage-mounted mooring rings or mooring rings mounted above the landing gear shock struts will have enough slack to prevent ground resonance. Chains attached to landing gear axle-mounted mooring rings will have no slack.

CAUTION

Engine and APU starts, blade spread, and rotor engagement will not be done when wind conditions exceed the limitations in the individual aircraft operator's manual.

NOTE:

1. After the helicopters are positioned for launch, ensure that they are moved as quickly as possible so that they do not exceed the APU run-time limitations during start and run-up.

2. An insufficient number of tow bars for CH-47s and ground-handling wheels for skid-equipped aircraft can cause delays in moving aircraft on deck. Aviation units should be prepared to bring as many tow bars and ground-handling wheels on board as possible.

(2) When possible, aircraft should be spotted for night amphibious operations from bow to stern in the event sequence shown in the air plan.

(3) The relative wind direction and velocity will be passed to the pilot by a prearranged method (5 MC, hand signal, or radio) before engine start, rotor blade spreading, or engagement.

(4) The LSE or director will receive clearance from the air officer before the engines are started or the rotors engaged.

c. *Wind and Deck Limitations.* Safe aircraft operations require strict adherence to prescribed wind and deck limitations. Commanding officers may establish more restrictive limitations.

NOTE: For specific wind and deck limitations, see the aircraft operator's manual or shipboard operating bulletin.

d. *APU or GPU Start.*

(1) When aircraft are spotted on the flight deck, pilots will proceed with the prestart procedures and signal the LSE or director when they are ready for the APU or GPU to be started.

(2) The LSE or director will request clearance for the APU or GPU start from the air officer in PriFly through the flight deck officer. PriFly will display a red rotating beacon and make the following announcement over the 5 MC: "Check chocks, tie-downs, fire bottles, and all loose gear about the flight deck; helmets buckled; goggles down; start APU/GPU on LSE/director signal."

(3) The LSE or director will relay the clearance to the pilot before the APU or GPU start can be initiated. After the APU or GPU is started, radios will be turned on and set to the land/launch frequency, when practicable. Approved shipboard fire-fighting equipment with the appropriate reach nozzles will be manned for APU, GPU, or main engine starts.

e. *Radio Check.* After an APU/engine start, PriFly will initiate a radio check. When several flights must respond, the order normally will be from bow to stern. Normally, an aircraft that fails to complete a successful radio check will not be launched. With the aviation unit commander's concurrence, the air officer may authorize the launch of an aircraft without UHF radio only when--

- Escort aircraft are provided.
- Two-way radio communications exist between aircraft in the flight.
- VMC exists for the planned route and length of the flight.

NOTE: These procedures may not work for some units, especially special operations units that have unique communications equipment. In these cases, the unit SOP should cover the communications check and the unit should coordinate these procedures with PriFly.

f. *Navigation Gear Alignment.*

(1) Because of the motion of the ship in three axes, the alignment of navigation equipment, especially the LDNS, may be impossible.

(2) Software updates for the OH-58D(I) allow the ship's course and velocity to be entered into the navigation computer before the equipment is aligned. Embedded GPS navigation systems allow constant updates of the aircraft's navigation system or position, velocity, and altitude. This allows the navigation equipment to be aligned while the ship is moving.

(3) For aircraft that are not equipped with the LDNS or GPS, the options for navigation system alignment include the following:

- Attempt alignment in flight and update your position over the ship.
- Initially align over the ship and update the system over a land-based point, if possible.
- Use the technique shown in Figure 4-7 to align navigation equipment on AH-64A aircraft without an embedded GPS.

1. Enter GPS present position in SP1.
2. HARS switch--fast for 2 to 3 minutes.
3. HARS switch--operate; ground crew--deactivate squat switch.*
4. CPG--enter ship's heading in SP1.
5. HARS should align within 90 seconds.
6. Ground crew--release squat switch.
7. HARS switch--OPR and doppler off until after takeoff.
8. After takeoff, doppler--On/Update doppler and HARS using GPS.

*Deck personnel must exercise extreme caution when moving around aircraft while the squat switch is deactivated.

NOTE 1: During alignment, the ship's heading must remain relatively constant; some degradation will occur after several takeoffs and landings.

NOTE 2: There may be times when the only way to align this system is to use the in-flight procedure. Again, extreme caution must be taken when doing this procedure during the hours of darkness. In addition, the unmodified 128-series doppler is unreliable over water. The 137-series doppler works reasonably well but must be updated every 20 to 30 minutes using a GPS. **The doppler provides no useful navigation information until the aircraft clears the deck.**

Figure 4-7. Technique for navigation equipment alignment (AH-64 without embedded GPS)

g. *Engine Start.* When the pilot is ready to start the engines, he requests clearance from the LSE or director by raising his hand and displaying one or two fingers to show which engine he wants to start. The LSE or director requests clearance from PriFly through the flight deck officer. PriFly ensures that the winds are within the limits for starts or engagements. PriFly also displays a red rotating beacon (amber for skid-configured helicopters) and announces over the 5 MC circuit clearance to start the engine. When the LSE or director gives the signal, the pilot starts the engines.

WARNING

Rotor brake failure is recognized as an emergency. Before the rotors are disengaged after a known or suspected rotor brake failure occurs, optimum winds will be provided for shutdown and the rotor blade system will stop. (Aircraft without rotor brakes may require a waiver to operate on Navy ships.)

NOTE: Weapons on aircraft racks or launchers will be latched mechanically before the engines on the aircraft are started.

h. *Rotor Engagement.*

(1) When the pilot is ready to engage the rotors, he gives the LSE the ready-to-engage signal. The LSE relays this request to the flight deck officer, who signals PriFly.

(2) Rotor blades should not be engaged while the ship is in a turn unless approved by the ship's CO or his designated representative. The anticipated winds and ship's heel will be communicated to the helicopter aircraft commander before the turn is executed.

(3) The air officer will ensure that proper wind conditions exist for engagement. If the winds are high, rotor engagements should start with the downwind aircraft and work upwind.

WARNING

1. Reported winds displayed in PriFly may vary greatly from the existing winds over the deck.

2. Use extreme care when engaging or disengaging rotors if other aircraft are being launched or recovered.

3. Not all Army helicopter wind envelops are found in Navy publications. Aviation units should be prepared to provide wind envelops to PriFly.

(4) When the pilot is ready to engage the rotors, an amber light is displayed to direct the flight deck officer and LSEs to give the engage signal to the pilot.

(5) Pilots of all aircraft will be provided the relative winds (both direction and velocity) using either radio, 5 MC, or hand and arm signals. Before attempting rotor engagement, each pilot will acknowledge clearance.

WARNING

1. Personnel will not walk under rotor blades until the blades have stopped or come up to full speed. Personnel must obtain clearance from the LSE before walking under rotor blades.

2. Personnel will not walk under the tail rotor of a single-rotor helicopter. During operations on FFG, DD, and DDG ships, walking under the main rotor and tail boom is the safest way to get around an armed helicopter.

i. Internal Cargo and Troops (Helicopter).

(1) Internal cargo normally is moved to the flight deck staging areas using cargo elevators near the island (LPH), fixed vehicle ramps (LHA), or aircraft elevators. The CCO or ATO directs loading. Internal loading varies according to the type of aircraft, cargo, and deck load. The CCO or ATO will ensure that pilots are notified of any changes to the prebriefed cargo loads.

(2) As directed by the CCO or ATO, combat cargo personnel (white shirts) escort troops to the flight deck by way of designated troop debark stations or shelters. Clearance will be requested from the LSE before loading or unloading troops while aircraft rotors are turning.

j. ***Downed Aircraft (DUDS).***

(1) The disposition of downed aircraft ("red X" on run-up) will be according to the prelaunch briefing. Except in an emergency, downed aircraft will be shut down only on signal from the LSE or director. Pilots will remain in downed aircraft until the crew chief or PC is ready to man the cockpit. (This does not apply to skid-configured helicopters.)

(2) After signal from the LSE or director, the downed aircraft will be shut down quickly. The maintenance officer or his representative will inform the flight deck officer of the nature of the trouble and give him the estimated repair time. If the required maintenance will take a long time, the aircraft normally will be put into the pack or taken below to the hangar deck. If repairs can be made on deck and succeeding launches will not be delayed, the aircraft will be launched to rejoin the flight when the maintenance officer places it in "UP" status.

Section III. Aircraft Launches

NOTE: Use of the aircraft handling signals in Appendixes A and B is mandatory when aircraft are being launched. Radio communications during launch, particularly between PriFly and armed helicopters, will be held to a minimum. The primary means of aircraft control on the deck will be hand and arm signals.

4-6. LAUNCH PROCEDURES

a. ***General.***

(1) When all prelaunch checks are completed and the pilots are ready to take off, they give the LSE or launch officer a thumbs-up signal and report to PriFly the aircraft status, fuel state, and souls on board. (PriFly may require that this information be reported after takeoff). The LSE or director signals the flight deck officer, and the flight deck officer notifies PriFly that all aircraft are ready for launch. PriFly requests a green deck from the bridge. When the ship is on a steady course, the OOD orders Hotel/Foxtrot flag close-up and gives PriFly a green-deck signal. The air officer ensures that the proper wind conditions exist for the launch according to aircraft operator manuals.

(2) Launching helicopters while the ship is in a turn should be attempted only after authorization from the ship's CO or his designated representative. The helicopter PC will be notified of the anticipated wind parameters and the ship's heel before the turn is executed.

(3) The air officer directs the flight deck officer to have chocks and tie-downs removed. The flight deck officer then directs the LSE or director to remove the tie-downs and chocks. Helicopter tie-downs will be removed from aft to forward. The main mount tie-downs will be removed simultaneously. When the LSE or director signals, each chockman (blue shirt) removes all the tie-downs and chocks from his side of the aircraft, goes to the LSE or director, and faces the pilot.

(4) Chockmen will carry the tie-downs so that the pilots can see them. They will show the pilots the tie-downs and the pilots will acknowledge them. The LSE or director will point to the chocks and tie-downs that were removed and show the pilot one finger for each tie-down that was removed. The pilot will indicate the number of tie-downs and chocks he sees and reply with a thumbs-up signal when he is ready to take off.

(5) When PriFly is satisfied that conditions are ready for a safe launch, the deck condition lights are set to green and the launch starts.

b. ***Ordnance-Equipped Aircraft.*** When an aircraft carrying ordnance needs to be armed, the launch officer or LSE directs the pilot's attention to the ordnance safety supervisor after he ensures that the aircraft is in the proper launch position and an initial walk-around has been completed. When arming is completed and the arming crew is clear of the aircraft, the ordnance safety supervisor gives the pilot a "thumbs up" signal and directs his attention back to the launch officer or LSE.

WARNING

Ordnance-equipped aircraft that are loaded with forward-firing ordnance normally are spotted on the flight deck. They are angled outboard in such a way that if an inadvertent firing occurred, the projectile would not hit the aircraft, flight deck personnel, or the ship's superstructure.

c. ***Helicopter Launch Procedures.*** When the green deck signal is given, the LSE rechecks to ensure that the aircraft is clear of all tie-downs and the area around the aircraft is clear of equipment and personnel. He also checks to ensure that all airborne aircraft are clear of the launch area. Only then does he give the pilot the signal to take off. The pilot will not start the takeoff until he receives the signal from the LSE and the wind condition from PriFly.

4-7. NIGHT LAUNCHES

a. Night launching procedures for helicopters are the same as for day, except--

(1) The LSE will use amber wands.

(2) Flight deck personnel will use goggles with clear lenses.

(3) During the prelaunch sequence, flight deck personnel, LSEs, and control tower personnel will cycle all control knobs and switches on the VLA control panels to ensure that each element is working properly. For those lights that may be obscured from the control panel operator's vision, an LSE will help confirm that the switch and knob settings are showing the right indications.

(4) SAR helicopters and rescue boats will be equipped with night-signaling devices.

(5) Pilots will ensure that all aircraft light switches are in the OFF position before the electrical system is activated.

(6) External aircraft light signals will be used as outlined in the night-lighting procedures shown in Figure 4-8.

(7) PriFly and the OOD will follow the maximum relative wind velocity charts shown in Appendix E for night helicopter launches and recoveries.

(8) Pilots should not initiate any radio frequency changes or heading changes before reaching 300 feet.

(9) Neither PriFly nor the HDC will require a frequency or heading change be made before the pilot reaches 300 feet unless doing so is required for safety reasons.

b. Night launches from spots that do not afford visual reference to the deck may be dangerous because visual cues are lost at lift-off.

HELICOPTER SIGNAL	HELICOPTER LIGHTS
Ready to start APP	Red cockpit dome light or red lens flashlight on
Ready to start engines	External navigation lights on STEADY DIM
Ready to engage rotors	External navigation lights on FLASHING DIM
Ready for takeoff	Anticollision lights on; navigation lights on STEADY BRIGHT
After takeoff	Anticollision lights on; navigation lights on STEADY BRIGHT
180° abeam position/right seat landing	Navigation lights on STEADY BRIGHT; anticollision lights on
180° abeam position/left seat landing	Navigation lights on FLASHING BRIGHT; anticollision lights on
After final landing or when on deck for extended period	Anticollision lights off; navigation lights on FLASHING DIM
Ready for disengage rotor	Red dome light or red lens flashlight on; navigation lights on FLASHING DIM
NOTE: May be modified by Pri-Fly to accommodate weather conditions and aircraft characteristics	

Figure 4-8. Night launch procedures

4-8. NIGHT VISION DEVICE CONSIDERATIONS

a. There are no standard Army procedures for using NVDs during shipboard operations. Aircrews must be keenly aware of the factors discussed below when they use NVDs in a shipboard environment.

(1) Pilots must be aware of the location of the deck, obstructions, and mechanical turbulence that the ship's superstructure induces.

(2) The prevailing wind and the ship's course and speed significantly affect relative wind and turbulence.

(3) Visual illusions, especially relative motion, and disorientation are important factors. The lack of visual cues and height perception problems increase a pilot's chances of disorientation. The pilot flying the aircraft should announce "vertigo" as soon as he becomes disoriented so that the other pilot can take the controls.

(4) During takeoffs and landings, the crew must be alert. The pilot flying the aircraft needs to keep his attention focused outside while the copilot assists in clearing the aircraft and monitoring system instruments.

(5) Determining the rate of closure is difficult because of the lack of references, especially when landing to the stern of a single-spot deck. If a safe landing is in question anytime during the approach, the pilot should perform a go-around.

(6) Some LPD and LPA ships have NVG-compatible lighting. However, the deck lights on most ships are not NVG-compatible. While the brightness of the lights can be dimmed considerably, they still may cause the NVG to wash out.

b. While aviators must understand certain procedures, the ship's company also must understand the factors discussed below before starting NVD operations.

(1) Flight deck lighting levels depend on ambient light levels. (Higher ambient light levels require higher light settings; lower ambient light levels require lower light settings and an overhead floodlight.) The use of red light should be reduced. Avoid the use of drop-line and overhead lights, wave-off lights, rotating beacons, and blue deck-edge lights.

(2) The NVD display is monochrome--all lights appear green. Color-coded light signals and landing aids are not useable.

(3) The SGSI should be off during NVD operations.

(4) The ship's navigation lights and their locations cause NVG wash-out. In low ambient light conditions consistent with safe navigation, consider turning off the stern light, the masthead light, and range lights while the ship is underway. While the ship is at anchor, turn of the stern anchor lights. While the ship is underway, side lights should be dim.

(5) LSEs should use NVG during launch and recovery operations. Other flight deck personnel do not need to use NVG.

(6) Aircraft deck moves should be stopped or kept to an absolute minimum during NVD operations.

(7) If tow tractors are used during NVD operations, their lights should be muted by taping the lenses.

(8) The air officer and/or OOD should have NVG available in PriFly.

(9) Landings should not be made in front of or between turning or parked aircraft.

(10) Other than LSE, flight deck personnel who must use lights should use green chemsticks or flashlights with approved blue lenses.

(11) Before starting NVG operations, a check should be made of darkened ship conditions.

(12) Aircraft that are in the landing pattern should use formation lights. Aircraft that are on deck for refueling or landing should use position lights on dim.

(13) The Army aviation unit should be prepared to provide ship personnel with NVG.

4-9. EMCON OR ZIP-LIP LAUNCH PROCEDURES

a. ***General Procedures.***

(1) When radio use is limited, other forms of communications may be employed. Visual communications become extremely important, including the proper use of the ship's flag command and display signals. Aircraft lighting, Aldis lamps, blinkers, and hand and arm signals are necessary to conduct flight operations safely. Both the aircraft and the controlling ship will monitor the land and launch frequency. Radio transmissions are not authorized unless they are required for safety of flight.

(2) All flight operations conducted under EMCON conditions will be briefed and coordinated between the Army and the ship's controlling agencies. During EMCON conditions, increased emphasis will be placed on safe operating procedures.

(3) In addition to those command and display signals shown in Figure 4-8, the following signal will be used for tie-down removal before launch during EMCON conditions--a momentary display of the green beacon from PriFly to the LSE (red, green, red).

b. ***Day Launch Procedures.*** During EMCON conditions, day launch procedures are conducted the same as normal operations with one exception. Visual signals are used to replace routine radio transmissions. Pilots will ensure that all equipment that emits radio electromagnetic energy is set according to the EMCON conditions established by the ship's CO. The LSE will position the crew chief so that he can relay wind direction and velocity to the pilots.

c. ***Zip-Lip Procedures.*** During Zip-Lip (radio listening silence) operations, launch procedures will be the same as during EMCON conditions. The appropriate hand, flag, and light signals will be used unless radio communications are needed for safety of flight.

d. ***EMCON Night Launch Procedures.*** Night launch procedures during EMCON conditions are conducted the same as normal night operations except that light signals are used to replace routine radio transmissions. All communications, navigation, and flight equipment that is not essential for safe night operations will be secured.

4-10. EMERGENCY AFTER LAUNCH

a. ***VMC.***

(1) If an emergency requires an immediate landing, the pilot will prepare to jettison external stores or internal cargo and dump fuel as necessary. This will lower the gross weight of the aircraft below the maximum allowable landing weight.

(2) The pilot will advise the tower of the nature of the emergency and his desires. The air officer will inform the bridge of the situation, direct the preparation of the deck, and give the pilot an expected BRC and an estimated Charlie time. The expected Charlie time will be based on the time needed to clear the deck and get an acceptable WOD for the recovery.

(3) The pilot will observe the progress of the turn into the wind and deck preparation. He will try to time his approach to avoid arriving at the deck too early.

b. ***Night and VMC.***

(1) If an aircraft has an emergency during departure and needs to land immediately, the departure controller will provide vectors until the aircraft is picked up by an approach or a final controller. Every effort must be made to keep the aircraft on the departure frequency until it is safely aboard. PriFly will be advised immediately of the emergency and given the control frequency.

(2) Aircraft emergencies that do not require immediate recovery will continue normal departure procedures while the departure controller gets close control. Once acquired, close control and positive radar hand-offs will be used until the emergency aircraft has been recovered or diverted.

c. ***Lost Communications During Departure.*** If communications are lost during departure–

- IFF squawk according to ship's procedures.
- Remain visual and return to the ship for visual recovery if in VMC.

4-11. HELICOPTER DEPARTURE PROCEDURES

NOTE:

1. Army aircraft normally are not equipped with TACAN navigation equipment. If TACAN is unavailable, onboard navigation equipment can be used to meet shipboard ATC requirements. (The 3-mile arc can be determined with a functioning doppler or GPS.)

2. One Army aviation unit performing shipboard operations procured and installed off-the-shelf modular TACAN receivers. They were procured using flying-hour resources and installed by contract maintenance personnel. Without TACAN, IMC recovery during shipboard operations may not be possible. TACAN installation requires an AWR from ATCOM or CECOM. Certification for IFR flight or intentional shipboard operations under IMC also will require additional testing and certification from CECOM.

3. Units must continue to coordinate with the ship's crew to meet ATC requirements.

a. ***Case I, VMC Departure to Rendezvous.*** This departure may be used when IMC is not anticipated during the departure and subsequent rendezvous. Helicopters will clear the control zone at or below 300 feet or as directed by PriFly. Rendezvous will be accomplished at briefed points according to unit tactical doctrine.

b. ***Case II, VMC to VMC on Top.*** Weather at the ship will not be less than 500 feet ceiling and one mile visibility. Helicopters will depart by way of Case I departure and maintain flight integrity below the clouds. Weather conditions permitting, departure on assigned missions also will comply with Case I procedure. If VMC cannot be maintained, pilots will proceed according to Case III departures.

c. ***Case III, IMC/Night.*** Whenever weather conditions at the ship are below Case II minimums or when directed by the CO or OTC, helicopters will be launched at no less than 1-minute intervals, climb straight ahead to 500 feet, and intercept the 3-mile arc. They will arc at 3 miles to intercept the assigned departure radials. Upon reaching the assigned departure radial, they will turn outbound and begin the climb to the assigned altitude. Departure radials will be separated by a minimum of 20 degrees.

NOTE: When the only restriction to VMC is a lack of a visible horizon, modifications to Case III requirements may include, but are not limited to, night VFR operations near the ship, tactical troop lifts within the AOA, ship-to-ship movement, and NVG operations.

(1) Helicopters will be launched on the assigned departure frequency, vice land and launch, and monitor guard. PriFly will monitor the departure frequency once the helicopter is airborne.

(2) Helicopters launching on tactical missions will rendezvous as briefed and report KILO (aircraft mission readiness). They will then be switched to the assigned tactical control agency.

(3) Departing aircraft will report--

- Airborne.
- Arcing (if applicable).
- Established on departure radial.
- Popeye with altitude.

NOTE: When in IMC, Popeye is a mandatory report for single aircraft when they reach the assigned departure altitude. This report alerts the departure controller that further instructions are required.

- On top with altitude.
- KILO (mandatory).

(4) Minimum separation for departure radials is 20 degrees. If applicable, assignment depends upon--

- The mission of the aircraft.
- Topographical features.
- Reserved radials for emergency use.
- Ships in formation.
- Airspace restrictions (ADIZ, hot, warning, restricted, and prohibited areas).

4-12. CONTROL OF DEPARTING AIRCRAFT

The pilot is primarily responsible for adhering to the assigned departure. However, advisory control normally will be exercised with a shift to close control when required by weather conditions, upon request, or when the assigned departure is not being followed. After launch, the AOCC or HDC will--

a. Record data on the status boards as required.

b. Ensure that communications and positive track are established with all aircraft to the extent possible under existing EMCON conditions.

c. Request NAVAID checks as necessary.

d. Maintain advisory control of departing point-to-point flights until the pilots shift to en route frequencies.

e. Maintain advisory control of other aircraft until control is accepted by CIC, TACC, or another controlling agency.

f. Before releasing aircraft to another controlling agency, give each pilot (or flight leader) any pertinent information such as changes in the composition of the flight, PIM, or mission.

g. When transferring control of an aircraft, give the new controlling agency the distance and bearing of the aircraft being transferred and obtain acknowledgment of the assumption of control.

h. File flight plans as necessary.

CHAPTER 5

AIRCRAFT RECOVERY

This chapter contains information on approaches and recoveries to amphibious ships (LPH, LHA, and LHD) and aircraft carriers (CV and CVN) after missions. It also covers safety considerations; single-spot ship operations are covered in ***Appendix D.***

Section I. Arrival

5-1. PROCEDURES

Once released by mission controllers, aircraft entering the ship's control area will switch to the AOCC/HDC frequency for further clearance to the marshal pattern. Adjustments to the landing order may be made to accommodate aircraft materiel conditions, fuel state, and hung ordnance. Flights will check in with AOCC/HDC when they enter the control area or when directed by other control agencies. These aircraft will provide the following information:

- The aircraft call sign.
- The position relative to the ship.
- Altitude.
- Fuel state (hours and minutes of lowest fuel state in flight).
- Souls on board.

AOCC/HDC will respond with the following information:

- The expected approach time.
- Marshal instructions, if required.
- Vectors, if required.
- The estimated recovery time.
- The altimeter setting, wind, weather, and BRC.
- The time.
- Other pertinent information.
- Clearance into the control area.

Pilots will report "See you" when they make visual contact with the ship. AOCC/HDC will switch the flight to the PriFly frequency at 5 nautical miles (VMC).

a. ***Case I, Visual Descent or Approach (Helicopter).*** (See Figures 5-1 on page 5-3, 5-2 on page 5-4, 5-3 on page 5-5, and 5-4 on page 5-6.) Case I may be used when flights are not expected to encounter IMC anytime during the descent, break, and final approach. Weather minimums of 1,000 feet ceiling and 3 miles visibility are required in the control zone.

NOTE: During mixed aircraft operations, helicopters will enter the starboard Delta pattern which is located one mile to the starboard side of the ship at 300 feet and oriented on the BRC. (See Figure 5-1.) During mixed aircraft operations, the helicopter break altitude will not exceed 300 feet.

b. ***Case II, Controlled Descent or Visual Approach (Helicopter).*** Case II will be used during the day when weather conditions are such that flights encounter IMC during the descent but VMC of at least a 500-foot ceiling and 1-mile visibility exists at the ship. Positive control will be used until the flight leader or pilot reports that he has the ship in sight. AOCC/HDC will be fully manned and ready to assume control of Case III recoveries if the weather deteriorates below Case II minimums.

NOTE: Case II approaches will not be flown when Case III departures are in progress. Case III approaches will be used during marginal VMC.

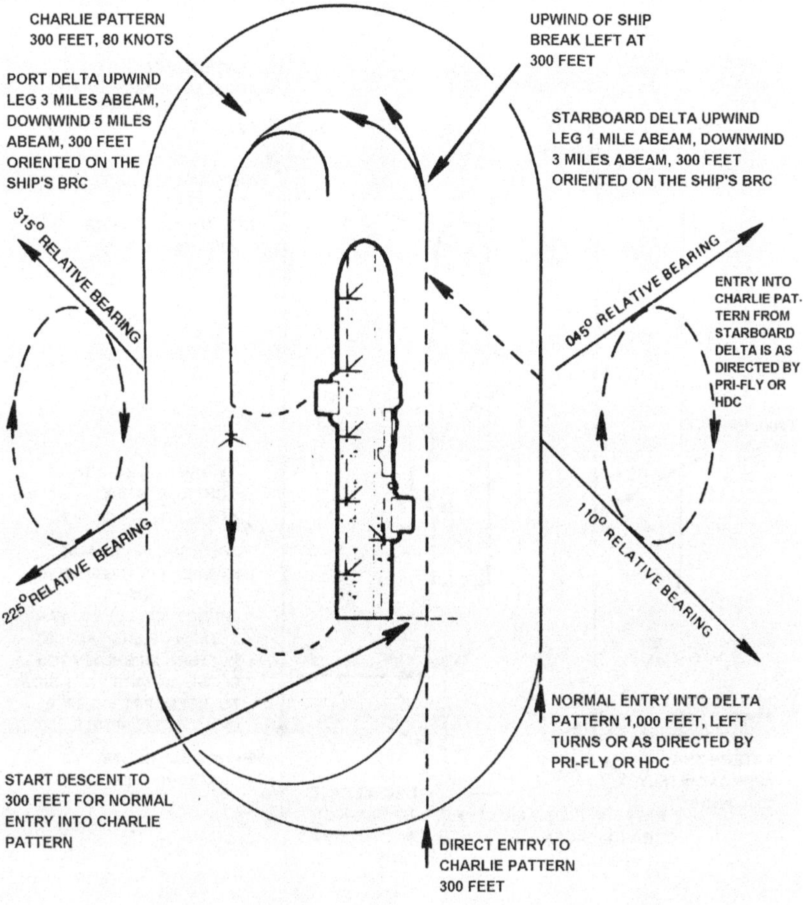

Figure 5-1. Delta and Charlie patterns for helicopters

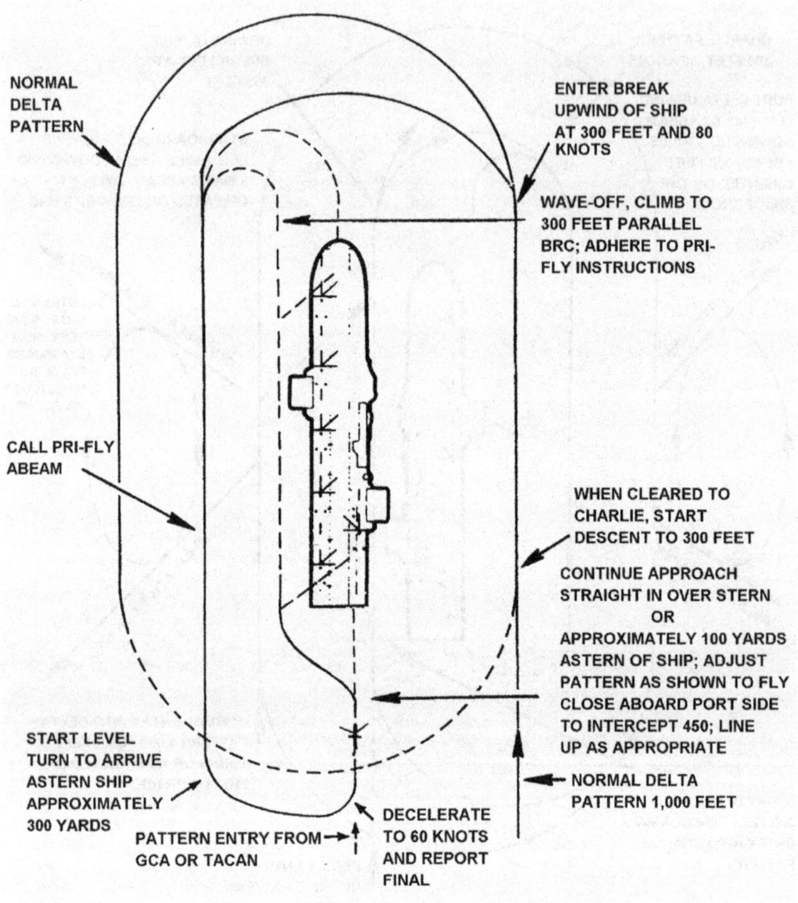

Figure 5-2. Helicopter night Case I recovery pattern

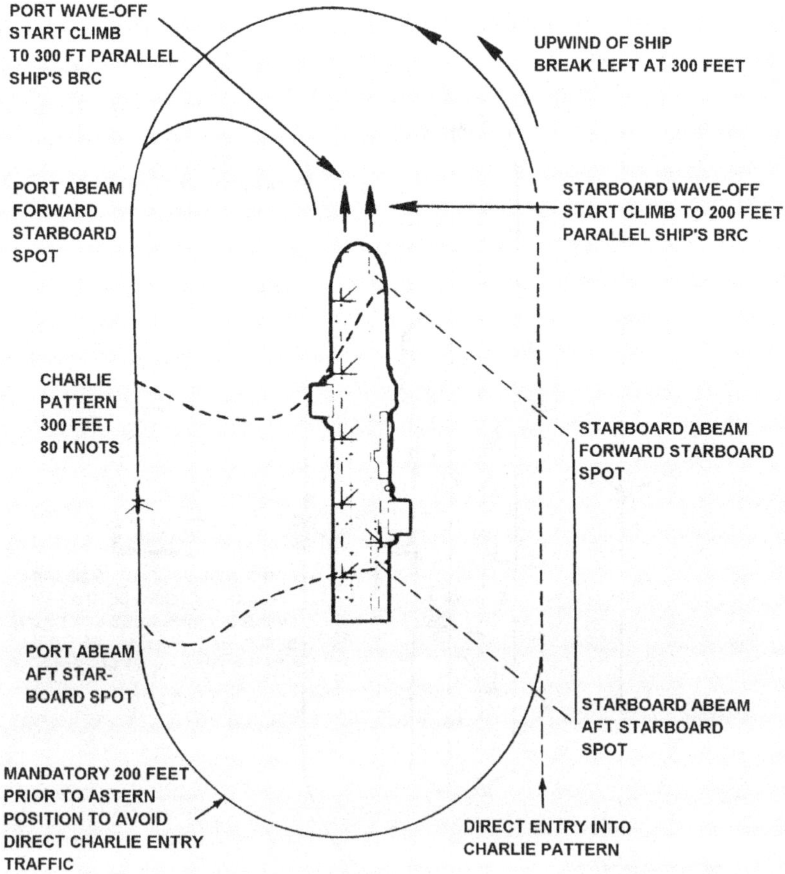

Figure 5-3. Helicopter recovery patterns for starboard side spots

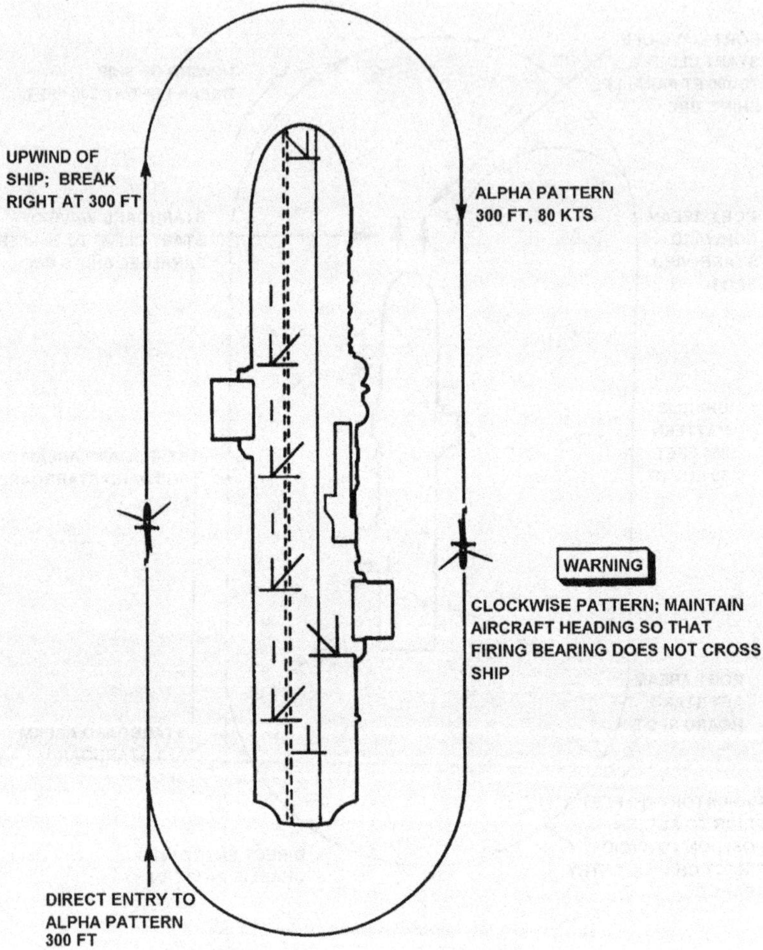

Figure 5-4. Alpha pattern--recovery of armed helicopters

c. ***Case III (Helicopter).*** (See Figure 5-5 on page 5-8.) Case III will be used when weather conditions at the ship are below Case II minimums. Unless modified by the ship's CO or OTC, Case III also will apply to night flight operations when there is no visible horizon. Case III formation recoveries are not authorized unless an aircraft experiencing difficulties is recovered on the wing of another aircraft. A straight-in, single-frequency approach will be provided in all cases. Formation flights by dissimilar aircraft will not be attempted except in extreme circumstances when no safer recovery methods are available. Precision radar will be used whenever available. The following procedures are mandatory for all Case III recoveries.

(1) ***LPH, LHA, and LHD marshal patterns.*** The assignment of marshal patterns depends on the topographical features, ships in formation, operational restrictions, and aircraft capabilities. The sky should be clear. While the following procedures are written for TACAN-equipped aircraft, these procedures can be adapted using other Army-specific navigation equipment (Doppler, GPS, INS, and so forth).

WARNING

TACAN marshal two will not be used during mixed aircraft operations.

NOTE: All bearings are relative to the BRC. All legs are two nautical miles long, standard rate turns.

- **TACAN marshal one**--180-degree bearing at 7 miles, altitude as assigned.
- **TACAN marshal two**--270-degree bearing at 7 miles, altitude as assigned.
- **TACAN marshal three**--090-degree bearing at 7 miles, altitude as assigned.

WARNING

Base altitudes for TACAN marshal patterns one, two, and three will not be less than 1,000 feet.

- **NDB/TACAN overhead marshal**--Overhead holding pattern on the 30-degree relative bearing, altitude as assigned (not less than 500 feet), 1 minute and a two nautical mile racetrack pattern, left-hand turns. See Figure 5-6 on page 5-9.)

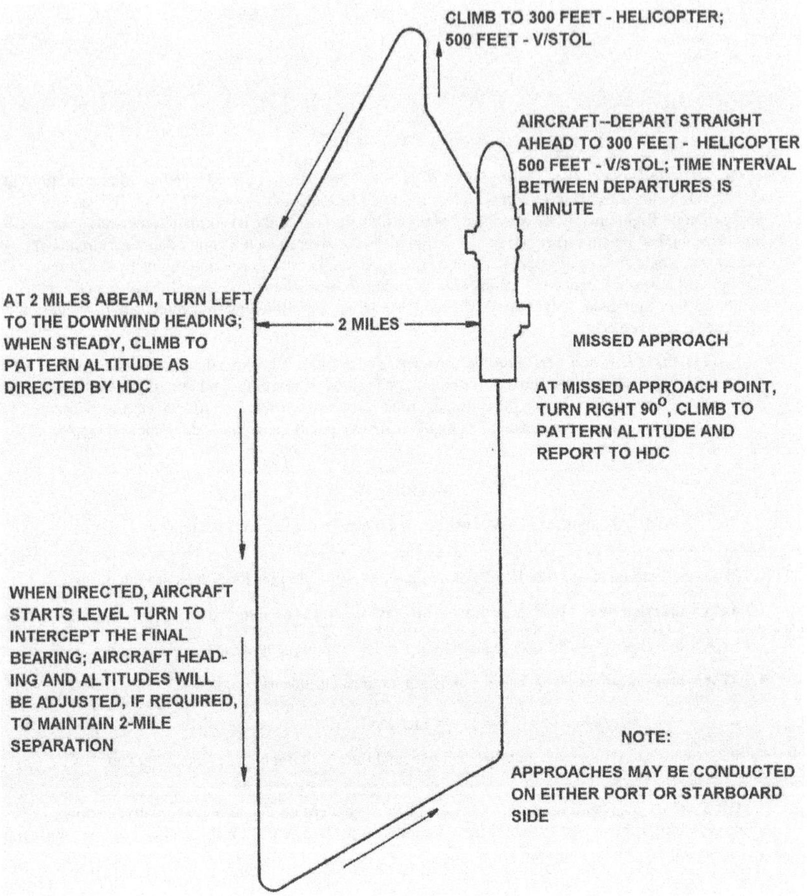

Figure 5-5. Case III carrier qualification pattern

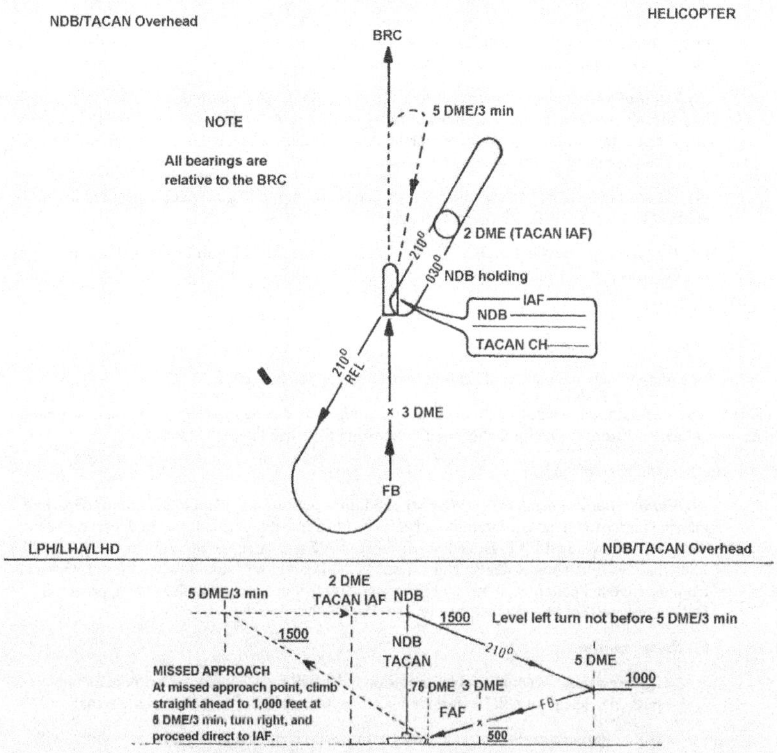

LANDING MINIMUMS				FLIGHT DECK ELEVATION	HIGHEST OBSTRUCTION
CATEGORY	LHA	LPH			
S-PAR	270-1/2	250-1/2	200(200-1/2)	LPH-50'	LPH-150'
S-ASR S-NDB/TAC	370-3/4	350-3/4	300(300-3/4)	LHA-70'	LHA-200'

Figure 5-6. Approach chart (LPH, LHA, LHD NDB overhead)

(2) ***CV and CVN marshal patterns.*** Aircraft carrier marshal patterns differ from those for amphibious assault ships in the way they are numbered and the altitudes at which they are flown. The CV/CVN marshal stack starts at altitudes of at least 5,000 feet and at 20 DME on the 180 degree radial. Because these marshals are not practical for helicopters, the Army aviation unit and the ship's leadership must agree on a marshal stack altitude during the presail conference.

(3) ***Marshal altitude assignment.*** Altitudes in marshaling will be assigned in VMC, if possible. If one aircraft is having communications or navigation equipment difficulties, a formation of two aircraft may be assigned to the same marshal for a section approach. Otherwise aircraft will be separated vertically by a minimum 500 feet.

(4) ***Marshal airspeeds.*** Marshal airspeed will be based on holding airspeeds specified in the applicable aircraft NATOPS flight manuals.

(5) ***Approach instructions.*** AOCC/HDC will provide the following information to each aircraft before approach clearance is issued:

- The new EAT.
- The final control frequency.
- The type of approach and outbound bearing (overhead approaches only).

NOTE: The assigned outbound bearing will be updated continuously during recovery to maintain a minimum of 20 degrees clockwise from the reciprocal of the final bearing (overhead approaches only).

- Frequency and IFF changes.

(6) ***Marshal pattern departures.*** Weather conditions permitting, operational aircraft departing the marshal pattern will have a 1-minute separation. Pilots must adjust patterns to depart the marshal pattern at the assigned EAT. Deviations from the EAT will be reported to the marshal controller immediately so that adjustments can be made to the interval for the following aircraft. Descents from the marshal pattern will be at 90 knots and 500 feet per minute to the final approach fix. Helicopters will assume the landing configuration before passing FAF.

(7) ***Radar approach.***

(a) ***Precision.*** When available, precision radar will be used to the maximum extent possible. The pilot will be provided heading and glide slope information on final.

(b) ***Nonprecision.*** When glide slope information is not available, aircraft on final will continue the descent to MDA after passing FAF. The final controller will provide the pilot with recommended altitudes and enough information to maintain accurate azimuth and safe altitudes until the aircraft reaches nonprecision minimums.

(8) ***Approach minimums.*** Approach minimums are shown on the ship's approach charts. The CO may increase these minimums if required by significant changes in operational capability such as decreased AOCC/HDC or proficiency of the embarked unit..

(9) ***Missed approach wave-off.*** Helicopters executing missed approaches will turn right 90 degrees, intercept the 5-mile arc, and arc right to reenter on the final bearing. If the deck is fouled or there are an excessive number of wave-offs, AOCC/HDC will direct all aircrews to proceed according to the last clearance and stand by for new instructions.

(10) ***Divert field.*** When a suitable divert field is available, aircraft will not begin an approach if the reported weather at the ship is below minimums unless the aircraft has enough fuel to proceed to the divert field in the event a missed approach is required.

5-2. LOST COMMUNICATIONS OR NAVIGATION AIDS DURING THE APPROACH

If communications or navigational aids are lost during the approach, the procedures outlined below will be followed.

- During IFF, squawk according to ship's procedures.
- During VMC, remain VMC and continue the approach using VFR lost communication signals and procedures.
- During IMC or at night and only aircraft communications are lost, continue the approach. Try to contact the ship using the survival radio if time permits and safe aircraft control is not jeopardized. When visual contact with the ship is made, follow the signal procedures shown in Figure 5-7.

FROM AIRCRAFT TO SHIP	
PILOT'S DESIRES OR INTENTIONS	**SIGNAL**
1. I require immediate landing.	Fly by or hover close aboard starboard quarter, remaining clear or other traffic, with gear DOWN and floodlight/landing light ON. With complete electrical failure, fire a red flare on a safe bearing away frolm the ship (helicopter only).
2. I desire to land but can wait for the next recovery or scheduled recovery time.	Fly by or hover on the starboard side of the ship, low and close aboard with navigation lights BRIGHT and FLASHING and anticollision lights ON. With complete electrical failure, fire a red flare on a safe bearing away from the ship (helicopter only).
3. I am proceeding to the divert field.	Fly up the starboard side of the ship, rocking wings with landing gear UP, navigation lights BRIGHT and STEADY and anticollision lights ON. If fuel state and the nature of the emergency permit, continue making passes until joined by a wingman. Upon reaching divert fuel state, proceed alone, setting IFF to emergency when departing.

Note:

1. The requirement for gear down and rocking wings as the signal for an immediate landing is not required when the pilot considers it unsafe because of the nature of the emergency; for example, the loss of an engine in multiengine aircraft.
2. At night, aircraft flying close aboard the port side of the ship without lights are considered to have an emergency requiring an immediate landing.

Figure 5-7. Visual signals during EMCON or lost communications

a. ***Complete Communications or Navigation Failure.***

(1) The pilot of the signal aircraft may elect to continue the approach by dead reckoning to the MDA until at least 2 minutes have elapsed since the expected arrival time. The pilot will climb out on the final bearing until VMC is achieved or he reaches the emergency marshal altitude. He will fly the appropriate triangular pattern at altitude, conserve fuel, and expect join-up. The pilot will follow lead or, at his discretion, fly to a divert field fuel permitting. If below overcast, he will fly the DR search pattern to locate the ship. When visual contact with the ship is made, the pilot will follow the procedures shown in Figure 5-7.

(2) The pilot may elect to discontinue the approach. If so, he will climb on final bearing to VMC or emergency marshal altitude using the DR and follow the procedures shown in Figure 5-7.

NOTE: In the recovery procedure, an aircraft with inoperative navigation and/or communications equipment that is with or joined by an escort aircraft with operable navigation or communication equipment is handled as a single flight. The escort aircraft becomes the flight leader and normally communicates with the distressed aircraft according to standard procedures. The distressed aircraft assumes a position on the starboard wing of the lead aircraft. When the pilot of the lead aircraft has the ship in sight, he visually communicates a lead change. The distressed aircraft will complete a visual approach to a landing. The escort aircraft will enter Charlie pattern for a landing. If conditions preclude continued flight in the Charlie pattern, the escort aircraft will climb straight ahead on the BRC to one nautical mile DME or two minutes. Missed approach instructions will be complied with according to the applicable TACAN approach or as HDC/AOCC instructs.

b. ***Helicopter Emergency Marshal.*** An emergency marshal pattern provides an established procedure for aircraft experiencing lost communications to return to the ship during IMC. Pilots will be briefed on the emergency marshal pattern before initial takeoff. (These procedures presume that the TACAN azimuth and DME are operable.)

NOTE: TACAN marshal pattern 3 conflicts with the emergency marshal pattern.

(1) LPH, LHA, and LHD operations are unique in that helicopter final recovery times cannot be predicted because of mission status and the use of hot refueling. Therefore, an emergency marshal procedure must be established. This procedure will remain in effect throughout the aircraft event and will not have to be updated when the final recovery time of the aircraft is extended by hot refueling. The emergency marshal pattern for helicopters provides for the recovery of 24 individual helicopters experiencing lost communications or IMC.

(2) Each aircraft on the ship's air plan will be assigned an emergency marshal point. Radial, DME, EEAT, and altitude assignments are based on the assigned marshal point. The marshal point assignment will not be changed during the aircraft event except as requested by AOCC/HDC or the pilot and then only with the approval of both parties.

(3) The 24 marshal points are positioned on three TACAN radials and eight DME fixes at eight separate altitudes. The system provides lateral, vertical, and time separation. Radials are labeled A through C and are 45 degrees apart. Assigned radials are relative to the EEB.

(4) A helicopter experiencing lost communications during IMC will proceed outbound from the ship and climb or descend to the assigned emergency marshal altitude. It then proceeds directly to the assigned emergency marshal fix.

NOTE: During mixed operations, helicopters will cross at or above 2,000 feet.

(5) The appropriate IFF code will be squawked--mode 3, code 7700 for 1 minute, followed by mode 3, code 7000 for 14 minutes. The holding pattern is standard (right-hand turns) with the assigned DME being that point at which the outbound turn starts. The inbound turn starts at the DME fix plus two run. The pilot will maneuver the helicopter to be at the assigned DME fix on the assigned altitude at EEAT. At EEAT, he starts the descent to 500 feet and proceeds inbound to the five nautical mile DME arc. At the five nautical mile DME arc, the pilot turns left, and arcs clockwise to the EFB. He turns inbound on the EFB and starts the descent from 500 feet at the final approach fix (three nautical mile DME) to the minimum descent altitude. A missed approach will be according to the published TACAN approach.

NOTE: Helicopter airspeed throughout the emergency marshal pattern is 90 knots except during holding. During holding, maximum fuel conservation airspeeds will be observed.

(6) The emergency marshal pattern has two sets of EEATs. When 16 or fewer aircraft are launched, the pattern is repeated twice each hour. When more than 16 aircraft are launched, the pattern is repeated every hour.

c. ***Smoke Light Approach.***

(1) This approach is used when available equipment does not allow normal procedures to be used. It also will be used when the ship cannot be acquired visually using normal procedures and ditching is imminent. Both the CO and the PC must agree to attempt the procedure.

(2) The aircraft is positioned 2 miles astern the ship and proceeds inbound (180 degrees relative bearing to the BRC). The aircraft descends at the pilot's discretion to 40 feet and 40 knots. Smoke or matrix lights are dropped every 15 seconds (or another prearranged interval), and the pilot is kept informed of the number of smoke lights in the water. The pilot at the controls follows the smoke lights up the ship's wake, adjusting his closure rate until he can see the ship.

d. ***Proximity to Land Masses or Other Control Zones.*** Emergency marshal patterns are designed for blue water operations. Close proximity to land masses or control zones will require modification of emergency marshal procedures because exact conditions cannot be predicted. The ship's air operations officer must assign emergency marshal patterns that do not conflict with other aircraft, existing obstructions, or other patterns in use.

e. ***Emergency Approach Procedures.*** If an emergency condition exists or the aircraft does not have enough fuel to comply with the assigned emergency marshal procedures, the pilot will squawk the appropriate IFF code, climb or descend to 500 feet, proceed to and intercept the EFB at the 5 DME, and proceed inbound. Upon noting the arrival of an aircraft not under positive control, the AOCC/HDC will clear all aircraft from the anticipated route of flight of the distressed aircraft.

5-3. AIRCRAFT DIVERSION

If weather conditions are below Case II, particularly at night, a divert field or ship should be provided. The squadron commander and the air operations officer are jointly responsible for ensuring that aircraft performance data pertinent to diversion is available and understood by air control personnel.

a. ***Responsibilities.***

(1) The ship's CO decides whether to divert the aircraft.

(2) At night and during IMC, the air operations officer makes the appropriate recommendations to the CO as to which aircraft should be diverted or when the aircraft should not be diverted for flight safety. During VMC operations, the air officer makes these decisions.

(3) The LSO makes timely recommendations to the air officer regarding diversions based upon unsatisfactory pilot performance or landing conditions.

(4) The pilot will inform PriFly when he reaches bingo state without divert instructions.

(5) If practicable, the air operations officer determines the condition of NAVAIDs, communications, and lighting at the divert field before the first night or IMC recovery.

b. ***Planning Considerations.*** The following factors must be taken into consideration when aircraft diversion is being planned:

- Aircraft fuel state.
- Range and bearing to the divert field.
- Weather at the divert field (both current and forecast).
- Status and availability of the divert field for the type of aircraft.
- Available navigation assistance.
- Ordnance restrictions.
- Mechanical condition of the aircraft.
- Condition of the flight deck.

c. ***AOCC/HDC Responsibilities.*** AOCC/HDC will be alerted when an aircraft is approaching diversion state and be prepared to take control of the aircraft when the divert order is issued. AOCC/HDC will--

(1) Advise the pilot of the name of the divert field and its magnetic heading and distance.

(2) Advise the pilot to check that the gear is up (if applicable).

(3) Instruct the pilot to shift to the control frequency en route.

(4) Provide the pilot with the latest available en route weather, the altimeter setting at the divert field, and the position from which divert was made.

(5) If operations are being conducted outside an ADIZ boundary, provide the pilot with the necessary ADIZ information. Advise the appropriate GCI site of the diverting aircraft's departure point, ADIZ penetration point, time of penetration, altitude, estimated time en route, destination, and any other pertinent information.

(6) Maintain a radar plot and radio monitor on the diverting aircraft for as long as possible and/or retain positive control responsibility for the aircraft until positive radar hand-off to the GCI, ARTCC, or another appropriate controlling agency.

(7) File a divert flight plan with the appropriate controlling agency. Ensure that similar information is provided to the pertinent air defense agency should an ADIZ penetration be necessary.

(8) Receive an arrival report for the diverting aircraft.

d. ***Pilot Responsibilities.*** After the pilot of the diverted aircraft lands, he will notify the ship by immediate precedence message.

Section II. Recovery

5-4. PREPARATIONS

The following actions will be completed in preparation for aircraft recovery:

a. The OOD will ensure that preparations for flight quarters are completed according to this manual. Operational checks of PriFly equipment, flight deck lighting, and optical landing aids normally will be completed before the beginning of air operations; which usually consists of manning the necessary stations.

WARNING

H-53E aircraft create more rotor downwash than any other embarked helicopter. This downwash can cause damage to unsecured rotor blades; blow aircraft chocks, tie-down chains, and tow bars about the deck or overboard; and cause injury or death.

b. Safe aircraft recovery operations require strict adherence to wind and deck limitations. The ship's CO may establish more restrictive limitations.

(1) *Approaches.*

(a) ***Initial contact.*** Flight leaders and/or aircraft commanders will report "see you" when the ship is in sight. At that time, AOCC/HDC will switch the flight to the land/launch frequency for PriFly control. Unless otherwise cleared by PriFly, flights will proceed and hold in the Delta pattern. (See Figures 5-1, 5-2, and 5-3.) The flights will plan their descent, break to meet the designated recovery time, and maintain an orderly flow of traffic into Charlie pattern.

(b) ***Standard Delta pattern.*** The Delta pattern is a VFR holding pattern established near the ship. Normally, the Delta pattern is a left-hand racetrack pattern around the ship. As shown in Figures 5-1, 5-2, and 5-3, it is oriented on the ship's heading, is close aboard the starboard side, and is flown at an optimum airspeed. During heavy traffic periods, additional Delta patterns may be used as assigned by PriFly.

(c) ***Helicopter starboard Delta.*** Normally, a holding pattern on the right side of the ship is a right-hand racetrack pattern that is oriented on the ship's heading at an assigned altitude (Figure 5-1).

(d) ***Charlie pattern.*** The Charlie pattern for all aircraft is a left-hand racetrack pattern on the port side of the ship. The upwind leg is a course that parallels the BRC. Unless otherwise directed by PriFly or AOCC/HDC, all aircraft will enter the Charlie pattern as shown in Figures 5-1, 5-2, and 5-3. The landing interval will be established or adjusted upwind so that it does not extend the downwind leg.

(e) ***Prep Charlie pattern.*** Aircraft cleared to Prep Charlie will conform to the normal Charlie pattern entry procedures. Once established in the pattern, they will conform to the racetrack pattern shown in Figures 5-1, 5-2, and 5-3 until PriFly clears them to land.

(f) ***Helicopter night Case I recovery pattern.*** The helicopter night Case I recovery pattern is a left-hand pattern on the port side of the ship. The pattern is extended downwind which allows the helicopter to complete the turn to final before starting the descent. The straight-in final approach is flown using the available visual landing aids such as SGSI, CAI Mod 2, and HAPI.

(2) ***Recoveries.***

(a) The air officer will keep the bridge informed as to the flight deck's readiness to land aircraft. When the deck is ready and the ship has settled on the recovery course, the CO or OOD gives PriFly clearance to recover aircraft. The air officer announces on the 5 MC, "Standby to recover aircraft spots 1, 3, and so on." A green rotating beacon will be displayed from PriFly, when appropriate.

(b) Helicopters should not be recovered while the ship is in a turn, except when authorized by the ship's CO or his designated representative. Information on the anticipated wind parameters and the ship's heel will be given to the aircraft commander before the turn is executed.

WARNING

A change in the ship's direction during recovery could result in a hazardous situation and put the helicopter outside recovery wind roll parameters.

(c) Under VMC, all flights returning to the ship will be directed by AOCC/HDC to contact PriFly at 5 nautical miles when the flight leader or aircraft commander reports that he has the ship in sight. Upon initial contact with PriFly, pilots will advise of their position relative to the bow; such as, 2 miles off port beam; 1 mile astern; 2 miles off starboard bow; and then give the fuel state. (See Figure 5-9.)

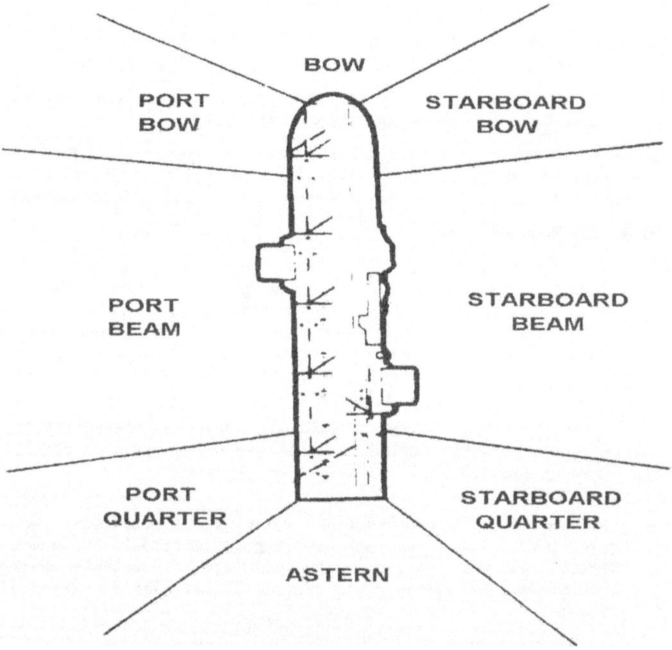

Figure 5-9. VFR relative position reporting

(d) The ship's land/launch frequency will be used for instructions and landing control. Strict radio discipline will be observed.

(e) Giving consideration to low-fuel-state aircraft, PriFly normally will give a "Charlie" to the number of aircraft for which there are available spots. Together with "Charlie," PriFly also will broadcast the BRC in degrees magnetic, altimeter, and wind condition across the deck. The pilot should plan the descent and break up to the landing pattern as shown in Figures 5-1, 5-2, and 5-3. Care should be taken to orient the landing pattern on the recovery course specified when it differs from the ship's heading. All pilots should take proper intervals at the upwind break.

(f) When a "Charlie number," such as Charlie 5, is given, the aircraft enters the landing pattern with the lead aircraft planning to be over the deck at the expiration of the number of minutes specified. A "Charlie" is given with the assumption that the first aircraft will be cleared to land upon arrival. When the "Charlie" spot number is given, the aircraft is clear to land. The pilot will indicate that the gear is down, if required, and give his seat position, as appropriate.

(g) LSEs will pick up landing helicopters at the 45-degree position in the approach turn of the Charlie pattern or at 100 yards astern in the helicopter night Case I recovery pattern.

NOTE: Wave-off and hold signals given by the LSE are mandatory; all other signals given by the LSE are advisory. Loss of visual contact with the LSE on final approach requires a wave-off.

(3) ***Landings.*** When aircraft are being recovered during VMC, the leader of the flight should plan to be on the deck with a minimum safe interval after the preceding aircraft has landed. When clearing aircraft to land, PriFly transmits relative direction, BRC, altimeter, and landing spots.

(4) ***Standard landing patterns***.

(a) The Charlie pattern is the standard Case I daytime helicopter landing pattern. The landing pattern for port spots is an approach that starts not later than when the aircraft is abeam the intended point of landing. A turn is then made to intercept the 45-degree line at the 90-degree position and flight continued straight into the spot. Helicopters should not be landed on spots directly in front of other helicopters.

WARNING

When helicopters approach on a 45-degree bearing to land directly in front of a spot occupied by another helicopter, rotor clearances (main and tail) between the two aircraft during the final portion of the 45-degree approach are significantly reduced.

(b) When approaching a spot directly in front of a spot occupied by another helicopter, the pilot should terminate the final portion of the approach on the 45-degree bearing at a point abeam the intended landing spot. From this point, the pilot flies the final transition by sliding sideways to a hover over the landing spot. The pilot in the right seat should land the aircraft.

(c) The Charlie pattern and the helicopter night Case I recovery pattern are the standard Case I night helicopter landing patterns. The air officer will ensure that all airborne aircraft and the squadron duty officer are informed when changing from one night landing pattern to another. Simultaneous use of the Charlie pattern and the night Case I recovery pattern is not authorized.

(d) Completion of the helicopter night Case I recovery pattern depends on the locations of aircraft on the landing spots. If the landing spots aft of the assigned landing spot are clear, the helicopter may complete a straight-in approach over the stern and air-taxi to the landing spot. If obstructions are between the stern and the landing spot, the air officer will direct the pilot to adjust his pattern to fly close aboard the port side and intercept the 45-degree lineup of the assigned landing spot.

(5) ***Nonstandard helicopter landing patterns.***

(a) ***Cross-deck.*** A cross-deck approach will be flown the same as the standard landing pattern except that the approach will continue across the flight deck to the assigned landing spot (Figure 5-3).

(b) ***Helicopter around stern.*** Starboard spots may be used by entering the normal Charlie pattern, calling abeam port quarters, and descending to 200 feet by the astern position. Continue up the starboard side to intercept an approximate 45-degree angle to the spot, and fly straight in (Figure 5-3).

(c) ***Helicopter modified straight-in.*** Depending on the amount of traffic in the pattern, PriFly may approve a straight-in approach to the spot.

NOTE: When directed, begin a straight-in approach far enough astern for the aircraft to be established positively on glide slope, at the proper airspeed, at a minimum distance of 1.5 miles, and at an altitude of 400 feet.

(6) ***Wave-off.*** Aircraft will be waived off --

- On voice command from PriFly or loss of communications with PriFly or the LSO.
- On command from the LSE or LSO.
- Any time the pilot feels that the approach cannot be completed safely.
- If visual contact with the LSE is lost on final approach (helicopters only).

The pilot will call "(aircraft identification) waving off" when the aircraft is parallel the BRC on the appropriate side of the ship and reenter the appropriate VMC recovery. Should reentry into the VMC pattern not be possible, the pilot will climb out straight ahead and request instructions from PriFly.

5-5. RECOVERY WITH ORDNANCE

a. ***In-Flight Procedures.*** Before the aircraft enter the ship's control zone, pilots will complete the following actions:

(1) Determine if all ordnance has been expended after the firing mission has been completed. Make a visual check between aircraft of all rocket pods.

(2) Every effort will be made to fire or jettison hung ordnance, as appropriate. If the hung ordnance cannot be fired or jettisoned, the pilot should consider diverting to a land base.

(3) Notify the ship as soon as possible if the ordnance must be brought back to the ship. Do not bring hung ordnance into the ship's control zone without clearance from the AOCC/HDC or the tower. Include in the initial notification the amount and type of hung ordnance.

(4) Properly safe all weapon systems.

b. ***Shipboard Procedures.*** Before the aircraft lands, the following actions will be completed:

- The bridge and other appropriate stations will be notified.
- The appropriate HERO condition will be set.
- Dearming crews will be standing by on station.
- Approval from the ship's CO will be obtained before ordnance is jettisoned from the ship.

c. ***Air Officer Procedures***. Before the aircraft lands, the air officer must complete the following actions:

- Clear a landing spot for recovery.

• Before the aircraft is recovered, announce on the 5 MC: "Stand by to recover (state type of aircraft) with hung ordnance on spot. Hung ordnance is (state amount and type). All personnel remain well clear of the flight deck area."

• Ensure that the ordnance safety supervisor and the squadron dearming team are on station before recovery.

• Ensure that all aircraft on the flight deck and in the landing pattern have secured HF and/or FM transmitters, IFF, TACAN, and radar altimeters, as required.

WARNING

All flight deck personnel, including LSEs, will remain clear of the line of fire and/or danger area of an aircraft landing with hung ordnance. Only the minimum required personnel will remain near the landing area. The pilot will not leave the cockpit until all ordnance and weapon systems have been safed properly.

5-6. HELICOPTER RECOVERY TIE-DOWN PROCEDURES

With the aircraft commander's concurrence and when the LSE gives the signal, chocks and tie-downs will be applied after the aircraft lands. They will remain attached until the aircraft is ready to be launched. During short on-deck times, such as when troops and supplies are being rapidly loaded, the aircraft may be chocked only. Tie-downs will be installed according to the individual aircraft operator's manuals. Unless otherwise specified, tie-downs will be attached to mooring rings in the vicinity of the main landing gear first.

WARNING

Any maneuvering of the ship while rotors of aircraft on the deck are turning will be done so that the winds and deck motion are kept within the operating envelopes of the aircraft. The pilot of an aircraft on deck with rotors turning will be informed of an impending ship's turn.

a. *Personnel Debarkation.*

(1) Pilots of ramp-equipped helicopters will not lower the ramps to discharge passengers until the LSE gives the signal.

(2) When troops are off-loading, the LSE will not signal for the ramp until the CCO troop handlers are present and recoveries or launches are complete on adjacent spots. The CCO handlers will escort the troops from the flight deck to the troop shelters as directed by the CCO. Flight deck, flight crew, or CCO personnel will escort passengers to a safe area.

b. *Rotor Disengagement.*

(1) Before rotors are disengaged and/or the engines shut down, the LSE will ensure that the signal to disengage is received from the flight deck officer who in turn receives the signal from the air officer. The LSE will ensure that the wheels are chocked, rotors are clear of personnel, and tie-downs are installed properly.

NOTE: Landing gear, external auxiliary fuel tank, and ordnance safety pins will be inserted before rotors are disengaged and/or engines shut down.

(2) Rotor blades should not be disengaged while the ship is in a turn unless authorized by the ship's CO or his designated representative. The aircraft commander will be advised of the anticipated wind parameters and the ship's heel before a turn is executed.

(3) The pilot will not disengage the rotors until he receives the signal from the LSE.

(4) The air officer will ensure that the proper wind conditions exist, according to the applicable NATOPS manuals, for disengaging rotor blades.

(5) If high winds exist, rotors will be disengaged starting with the forward most aircraft and working aft.

WARNING

Reported winds as displayed in PriFly may vary greatly with existing winds over the deck. Extreme care should be exercised when engaging or disengaging rotors if other aircraft are launching or recovering. Rotor engagement will not be attempted unless the tie-down configuration is as stated in the aircraft NATOPS flight manual. Failure to comply with this requirement may induce ground resonance.

5-7. NIGHT RECOVERY

Night operations are among the most critical for both pilots and flight deck crews. The tempo of operations will be reduced compared to day operations. To enhance safety, the pilots and flight crew must handle the aircraft slowly and carefully. All concerned personnel will be indoctrinated in night operations procedures.

a. *Postflight Maintenance Inspections.* Postflight inspections will be performed the same way they are for day operations. However, a red-lens flashlight will be used.

WARNING

Maintenance on or postflight of any portion of an aircraft that extends over the edge of the deck of the ship is prohibited.

b. *Flight Deck Operations.* Flight deck operations at night may cause some confusion between pilots and crews and the directors. All personnel must clearly understand the signals. During night operations--

- The LSE will be equipped with amber wands.
- All flight deck personnel will use clear lens goggles.
- All optical landing aids and flight deck lighting will be checked for proper operation before recovery operations begin.
- Lighting is critical. Under amber or red flight deck lighting conditions, white lens flashlights will not be used.

c. *Wind Limitations.* Wind limitations shown in the aircraft operator's manual or the shipboard operating bulletin will be complied with for all night recoveries. Commanding officers may establish more restrictive limitations. Flight leaders or aircraft commanders will be provided the ship's BRC as early as possible so the flight can maneuver properly to enter the designated pattern.

d. *Helicopter Night Recovery.* Recovery procedures will be the same as for day operations except external lights will be used for signaling (see Figure 4-7).

e. *Helicopter Optical Landing Aids.*

(1) The SGSI is a helicopter optical landing aid. It allows a pilot to visually establish and maintain the proper glide slope for a safe approach (Figure 5-10). The visual acquisition range is approximately 3 nautical miles at night under favorable environmental conditions. The light is projected through a horizontal arc 40 degrees wide. It provides a single vertical arc of green light (1.5 degrees), amber light (1 degree), and red light (6.5 degrees). The color of the light tells the pilot if he is above (green), below (red), or on (amber) the proper glide slope. By adjusting the aircraft's altitude to keep the amber beam visible, the pilot can maintain a safe 3-degree glidepath to the transition zone. Aircraft executing an SGSI approach normally will intercept the glide slope at the 180-degree position at 300 feet altitude and one-half mile abeam the intended point of landing. The SGSI normally is used from initial acquisition to about the 45-degree position where a visual transition is made to flight deck cues.

(2) Before starting a night TACAN straight-in approach, the SGSI should be set at 180 degrees azimuth relative to the ship. Helicopters executing straight-in approaches normally should intercept the SGSI glide slope at approximately 1,180 feet at a three-nautical-mile run. In addition to other radar approach voice reports, the pilot will report glide slope acquisition to AOCC, HDC, or PriFly, as appropriate.

NOTE: The glide angle indicator light uses the same beam colors as the SGSI. However, with the glide angle indicator light, the amber beam means that the pilot is above the glidepath, the green beam on the glidepath, and the red beam below the glidepath.

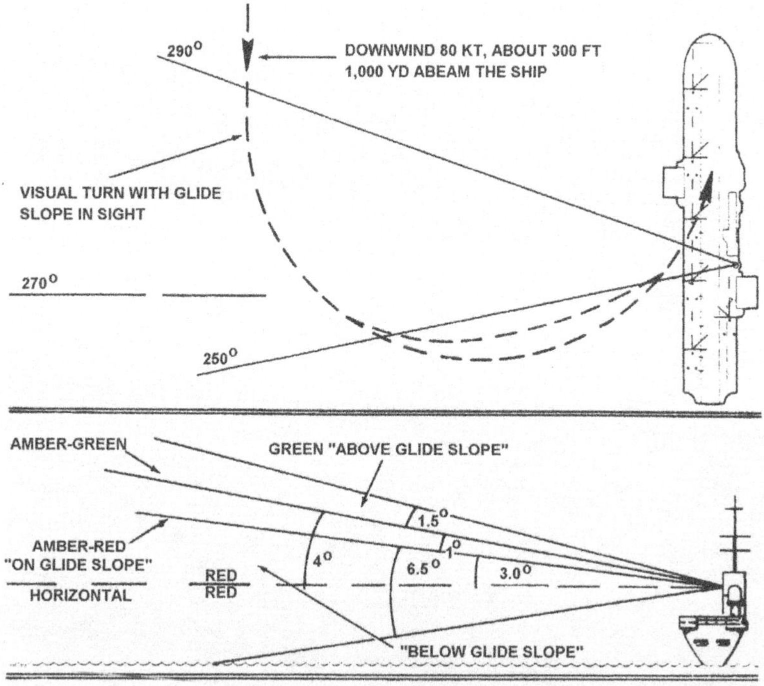

Distance (nm)	Height Above Water (feet)	
	Amber - Red (3.0°) LPH/LHA/LHD	Amber - Green (4.0°) LPH/LHA/LHD
3	1030/1050	1350/1370
2	712/732	925/945
1	393/413	500/520
1/2	234/254	287/307
1/4	155/175	181/201

Figure 5-10. Forward stabilized glide slope indicator

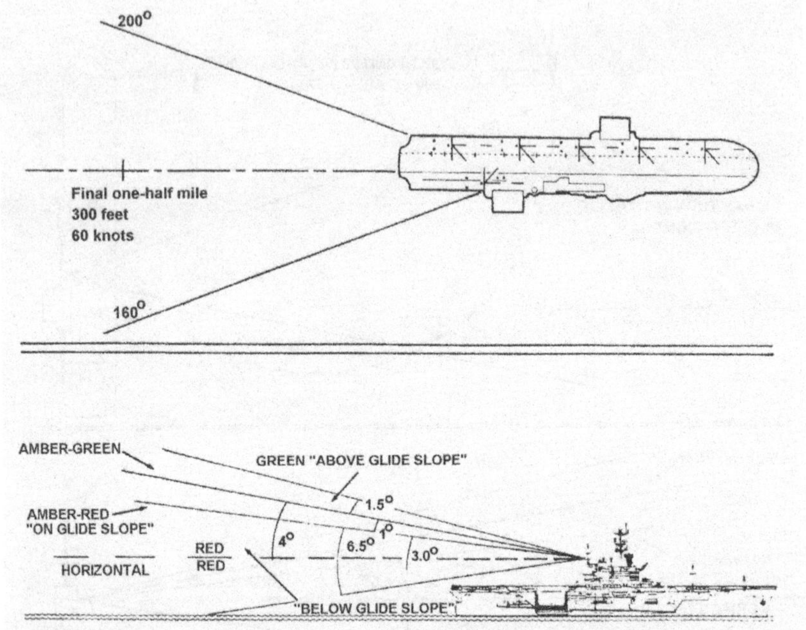

Distance (nm)	Height Above Water (feet)	
	Amber - Red (3.0°) LPH/LHA/LHD	Amber - Green (4.0°) LPH/LHA/LHD
3	1030/1050	1350/1370
2	712/732	925/945
1	393/413	500/520
1/2	234/254	287/307
1/4	155/175	181/201

Figure 5-10. Forward stabilized glide slope indicator (continued)

(3) A red wave-off light is installed on each side of the SGSI which is located forward on the island. The wave-off and cut lights of the CAI Mod 2 complement the SGSI located on the aft end of the island.

Section III. Safety

5-8. SPECIAL SAFETY PRECAUTIONS

The following items deserve special attention during aircraft operations:

a. While aircraft are being recovered, no personnel other than those required will be on the catwalks, flight deck, or elevators without the air officer's permission.

b. Personnel will not stand in or in any way block the entrances to the island structure or exits off the catwalks.

c. Personnel will not sleep or lounge on the flight deck, catwalks, galleries, or gun tubs during flight quarters.

d. Personnel will stay clear of all cargo elevator hatches and weapons mounts outlined by danger lines.

e. Personnel will not turn their backs to landing aircraft.

f. Except in an attempt to avert an accident, no director will give signals to a pilot who is being controlled by another director.

g. To reduce the possibility of an aircraft landing on a foul deck, landing spot and deck edge lights will not be turned on without the air officer's permission.

h. During instrument recoveries, PriFly will keep AOCC/HDC advised of the status of the deck and provide the estimated time the deck will be clear. AOCC/HDC will keep PriFly advised of the position of the nearest aircraft.

i. CIC and AOCC/HDC will keep PriFly informed of any aircraft known or suspected to have radio failure.

j. During night operations, green and red wands will be used only by the flight deck supervisor or launch officer.

k. Taking flash pictures during flight operations is prohibited.

l. Recovery will not be planned to any spot where aircraft or other obstacles would come within 10 feet of the rotor plane.

m. Left seat landings immediately behind a helicopter with the tail rotor turning are not authorized.

n. Night approaches to spot 1 are not authorized.

o. Left seat landings on spot 2 (LHA and LPH) are not recommended.

p. Right seat landings on spot 1 (LPH) and spot 3 (LHA) are not recommended.

5-9. EMISSION CONTROL, ZIP-LIP, AND LOST COMMUNICATION PROCEDURES

When radio communications are limited, operations may be conducted using other means of communication. Visual communications become extremely important, including the proper use of the ship's aircraft lighting, flag command, and display signals. The Aldis lamp, blinker, and hand and arm signals are necessary in conducting safe flight operations. These signals are explained in Appendixes A and B. Both the aircraft and the controlling ships will monitor the land/launch frequency. Radio transmissions will not be authorized unless required for safety of flight. All flight operations conducted under EMCON conditions will be briefed thoroughly and coordinated between the squadron and the ship's control-line agencies. During EMCON conditions, all personnel have an increased responsibility to conform to safe operating procedures.

a. ***EMCON Recovery Procedures.***

(1) The ceiling will be 500 feet above the highest normally prescribed Delta pattern with a minimum of 3 nautical miles visibility and a well-defined horizon. Returning pilots will plan to be in the Delta pattern before the scheduled recovery time. They will shift to and monitor the PriFly frequency when the ship is in sight. Each aircraft will have its anticollision lights on. Position lights will be on steady bright when the aircraft is within 10 nautical miles of the ship.

(2) Once established in the Delta pattern, the position lights will be set to flashing. The pilot will receive a flashing green Aldis lamp signal at the abeam position in the Delta pattern. The pilot will acknowledge by turning the navigation lights to steady-bright, leaving the anticollision light on, and descending to the Charlie pattern. At the abeam position, the pilot will receive a steady green Aldis lamp signal, conform to normal lighting procedures, and continue with the approach.

b. ***Zip-Lip Procedures.*** During zip-lip operations, recovery procedures will be the same as during EMCON. Unless radio communications are required for safety of flight, the appropriate hand, flag, and light signals will be used..

CHAPTER 6

AIRCRAFT AND WEAPONS HANDLING

This chapter provides general information on the handling of aircraft, weapons, and ammunition on Navy and Coast Guard ships.

Section I. Overview

6-1. GENERAL REQUIREMENTS

a. Designated aircraft directors control all aircraft movement. Aircraft will be moved only with the authority of the aircraft handling officer, flight deck officer, or hangar deck officer. Aircraft handling personnel will report to higher headquarters any unsafe practices or any condition that may affect the safety of personnel or equipment.

b. When the ship is at flight quarters, the OOD ensures that all anticipated turns are passed to PriFly so they may be announced over the flight and hangar deck announcing systems as appropriate.

c. The aircraft handling officer will begin a respot early enough to avoid unnecessary haste. When aircraft are airborne, however, the value of keeping a ready deck for as long as possible should be considered. Deck stability, prevailing winds, weather conditions, and nonskid conditions govern the tempo of aircraft movements. The safety of personnel is the primary consideration.

WARNING

During arming and dearming of ordnance, the area ahead, behind and/or surrounding the aircraft must be clear and remain clear until arming or dearming is complete.

d. Communications incident to aircraft handling will be according to the existing EMCON condition.

6-2. SAFETY BRIEFING

Before any major respot, the aircraft handling officer will brief the flight deck officer, hangar deck officer, and other key aircraft handling personnel. This briefing will include the expected wind and deck conditions and any other safety information.

6-3. MAINTENANCE LIAISON OFFICER

a. The maintenance liaison officer will ensure that the aircraft handling officer is kept apprised of the aircraft status and maintenance requirements. He also will maintain liaison between the air department and unit line and maintenance personnel. For this purpose, aircraft status and maintenance request boards will be kept in flight deck control.

b. A maintenance liaison officer or his representative should be on duty at all times during flight quarters or general quarters. His normal station is flight deck control; however, he is free to move about the flight deck and hangar deck as necessary. Changes in aircraft status will be submitted to the squadron maintenance liaison officer and entered on the aircraft status board. Entries and changes to the maintenance request board will be handled the same way. To help the maintenance liaison officer, the unit maintenance officer will provide an aircraft status report. This report will include up aircraft, down aircraft, their estimated time in maintenance, special maintenance requirements, and information of interest to the aircraft handling officer. The aircraft status report will be updated--

- Before scheduled flight quarters.
- As early as possible during general quarters and unscheduled flight quarters.
- As changes occur.
- To reflect the status of recovered aircraft.

c. The maintenance liaison officer is responsible for the overall performance of crew chiefs, PCs, and troubleshooters. He will ensure that no aircraft is placed on jacks or is otherwise immobilized without permission from the aircraft handling officer. He also will obtain permission for APU, engine, and rotor maintenance run-ups. Maintenance functions involving electronic emission are limited by existing EMCON conditions.

6-4. EQUIPMENT

a. The flight officer will ensure that all tractors, spotting dollies, twin-agent units, tow bars, chocks, and other equipment used on the flight deck are in satisfactory condition and are used properly. The hangar deck officer has a similar responsibility regarding the equipment used on the hangar deck. All aviation support equipment operators will be licensed according to current directives. Tractor drivers will not operate tractors with defective brakes or steering. Discrepancies will be reported immediately to a competent authority. Defective tow bars, chocks, wheels, and tie-downs will be taken out of service and turned in for repair. Tow bars, chocks, and tie-downs not in use will be stowed in designated spaces.

b. Specific requirements for crash and salvage crews and equipment operator requirements are found in the US Navy Aircraft Firefighting and Rescue Manual and the US Navy Aircraft Crash and Salvage Operations Manual.

Section II. Aircraft Handling

6-5. AIRCRAFT MOVEMENT

The minimum deck crew for aircraft movement on the flight deck or hangar deck is two safety observers, a qualified plane director, and two chock handlers or tie-down men. With AH-1, UH-1, or OH-58D(I) helicopters, one handler also must be on the tail skid. A pilot, plane captain, or qualified brake rider will man the cockpit. The duties and safety rules for moving aircraft on flight decks and hangar decks are discussed below.

> **WARNING**
>
> When heavy weather conditions are forecast, as many aircraft as possible will be moved to the hangar deck and all aircraft will be secured.

a. ***Duties.***

(1) *Plane director.* When preparing to move an aircraft (towing by hand), the director ensures that--

• The cockpit is manned by a qualified brake rider.

• All unnecessary personnel are away from the aircraft.

• Only qualified personnel pump up the ground handling wheels for skid aircraft.

• The tow bar is securely attached to the aircraft and to the tractor. If the aircraft is to be moved by hand, the tow bar must be properly tended by another director or designated tow bar man.

WARNING

1. Towing helicopters while the rotors are engaged is prohibited.

2. When the flight deck is slick with moisture, do not attempt to hand push the aircraft if the pitch of the ship is more than 10 degrees and the roll is more than 5 degrees.

CAUTION

Hand pushing aircraft is inherently less safe than towing aircraft using a vehicle. Pushing should be used only as a last resort or because of operational necessity. Hand pushing becomes more dangerous as the roll and pitch of the ship increases.

• All chocks, tie-downs, power cables, and other servicing and securing devices are removed before moving the aircraft.

CAUTION

1. Tie-downs and chocks will not be removed before the tow bar is attached to the tractor.

2. When moving aircraft by hand, chocks and tie-downs will not be removed until all positions are manned, the brakes are checked firm, and the deck pitch is safe.

• The ordnance safety supervisor indicates that the aircraft is safe to move if weapons loading or downloading is in progress.

• Adequate clearance exists to permit safe movement of aircraft.

• Safety men are posted as required to clear the aircraft if bulkheads, obstructions, or other aircraft are nearby.

• The qualified brake rider indicates that the aircraft brakes have been checked, adequate braking pressure is available, and the brakes appear to be in working order.

• All personnel except those necessary for the move are clear of the aircraft.

(2) *Brake rider.* When manning the cockpit of an aircraft to be moved, the brake rider will--

• Ensure that the ejection seat safety pins are installed, and safety pins are in place in the landing gear and auxiliary tanks, as appropriate.

• Ensure that the seat and the rudder pedals are adjusted so that the brakes can be applied and the director can be seen at the same time.

• Ensure that the windshield and side panels are clear of grease, cleaning compound, or any other film that might limit visibility.

• Open the cockpit canopy, windows, or overhead hatches if conditions permit.

CAUTION

Deck winds over 40 knots require that cockpit canopies be closed, which prevents audible signals from passing between the brake rider and the director.

• Test the brake.

CAUTION

1. Aircraft brakes should be tested twice--once before the chocks are removed and again after the aircraft begins to roll.

2. Aircraft parking brakes will be released only on signal from the director.

• Advise the director of any unusual condition or aircraft discrepancy that might make movement hazardous.

• Use available safety equipment such as safety belts, shoulder harnesses, life preservers, and so forth.

b. ***Safety Rules.***

(1) Before the chocks and tie-downs are removed, the director will call for "brakes." The man in the cockpit will give verbal or visual confirmation that he is holding the brakes. The tail wheel and/or nose wheel of the aircraft will be unlocked only after the director gives the signal.

(a) Movement of aircraft will be slow enough to allow for a safe stop within the available clear space. Movement will never be faster than the chock handlers can walk.

(b) The director will ensure that he or another director is plainly visible to the brake rider at all times.

(c) Safety observers will be stationed as necessary to ensure safety clearance any time an aircraft passes near another aircraft, bulkhead, or other obstruction. Only directors or personnel designated by the flight deck officer or hangar deck officer will act as safety observers. The safety observer and the director in control of the aircraft will either have each other in sight at all times or place a second safety observer in position to relay signals.

CAUTION

Movement of aircraft will not be attempted if the sea state or maneuvering of the ship produces excessive motion. Should a maneuver become necessary that would result in excessive deck motion while an aircraft is being moved, an announcement of the impending turn will be made over the 1 MC, 3 MC, or 5 MC system in enough time so that the chocks and tie-downs can be applied before the turn starts.

(d) During high winds or when the deck is unsteady, chock handlers will tend each main wheel closely. The brake rider will apply partial brakes as necessary to prevent excess speed from building up. Under these conditions, aircraft will not be moved by hand except in extreme need.

(e) Aircraft will be moved by aircraft handling equipment unless deck space does not allow safe maneuvering of the equipment and towed aircraft. When moving aircraft by hand, the aircraft should be moved against the movement of the deck. Therefore, the aircraft always must be pushed rather than being allowed to roll with the movement of the ship.

(f) Tractor drivers will not move an aircraft except under the control of a director. If a director's signal is not completely understood, the driver should stop and await further instructions.

(g) Except in an emergency, tractor drivers must avoid sudden stops when towing aircraft.

(h) Directors, safety observers, and chock/tie-down handlers will be equipped with whistles that they will hold in their mouths while controlling aircraft movement.

(i) When an aircraft with inoperative brakes must be respotted, the cockpit will not be manned. Chock handlers will remain in position to chock the main wheels instantly if ordered. In addition, a deck crewman will be available with tie-downs ready.

(j) As the aircraft nears the parking spot, it will be slowed to a speed that will permit an immediate stop. Tractor drivers must watch the director and often cannot check clearance for themselves. Therefore, directors and safety observers must maintain safe clearance for the tractor during movement in close spaces.

(k) Before aircraft are backed into deck-edge spots, chock handlers will be in position to chock the main wheels instantly.

(l) Sometimes before an aircraft reaches an interim or final spot, the tow bar must be repositioned for a better path of movement. In this case, the aircraft should be chocked and the initial tie-downs installed before the tow bar is disconnected.

(m) When the signal is given, the brake rider will apply full brakes immediately. The brakes must be applied simultaneously, especially if the aircraft is being moved by hand. The brake signal is a sharp blast on the whistle accompanied by the standard visual signal.

(n) The main wheels will be chocked as soon as the aircraft stops. The director remains with the aircraft until the handling crew completes the initial four-point tie-down. The tractor is then unhitched, and the director notifies the brake rider that he may leave the cockpit. When practical, leave the tow bar attached to the aircraft. The crew chief or PC will inspect the tie-downs for the required number and proper installation.

(o) When aircraft are parked on the hangar deck, allow clearance for access to and operation of fog foam monitors and fire plugs and for the operation of hangar bay doors. Do not park aircraft, yellow gear, or any item in a way that prevents the engineering escape chutes on the hangar deck from being opened completely.

(p) Personnel will not ride on tractors except in the driver's seat.

(q) Chock handlers are not safety observers; safety observers are not chock handlers.

(2) ***Elevator operation.***

(a) Elevator operation will be coordinated with the maneuvering of the ship. Only qualified personnel will operate aircraft elevators. A director will supervise the elevator any time it is being raised or lowered. He must position himself so that he is in plain view of the elevator operator at all times. Elevators will not be operated without two-way communications (verbal or visual) between operators.

(b) Directors should position the aircraft on the elevator so that it can be towed directly off without having to be repositioned.

(c) Tie-downs and chocks will be set before elevator movement. Before signaling for the elevator to be raised or lowered, the director will check the safety stanchions for proper clearance then signal for the stanchions to be raised. The elevator operator will sound the warning horn; check to ensure that all personnel, aircraft, and equipment are clear; and raise the safety stanchions. When the stanchions are up, the director will signal for the elevator to be raised or lowered. The only time an elevator will be lowered when the safety stanchions are inoperative is for operational necessity. In this case, directors will be stationed near the elevator to warn approaching personnel. If the safety stanchions on the hangar deck should fail, a temporary lifeline will be rigged as quickly as possible. After the safety stanchions have been raised or the warning given, no person will attempt to board or leave the elevator.

(d) Elevators will remain at hangar deck level for as short a time as possible. An elevator carrying an aircraft to the hangar deck will not be lowered until a crew is standing by to remove the aircraft from the elevator when it reaches hangar deck level.

(3) ***Aircraft damage report.***

(a) Any damage to an aircraft, no matter how slight, will be reported to the aircraft handling officer, flight deck officer, or hangar deck officer immediately. One of these officers will report the incident to the air officer and inform the unit maintenance liaison representative. The aircraft will not be flown until authorized personnel inspect it and declare it to be in an "up" status.

(b) The flight deck officer and hangar deck officer maintain a record showing the director's name, model aircraft, and bureau number. The record also will contain a brief summary of the circumstances that resulted in aircraft damage, whatever the extent of the damage. Reports of these occurrences will be made according to OPNAVINST 3750.6 series.

(4) ***Aircraft security.*** Aircraft will be tied down as directed by the aircraft handling officer or his representative. Unless otherwise specified, only chain tie-downs will be used. Tie-downs will run from a proper tie-down fitting on the aircraft to a padeye on the deck. The tie-downs will not press against oleo struts, hydraulic lines, tires, or any other portion of the aircraft. When an aircraft is spotted next to an elevator, tie-downs will not be attached to the elevator or across the safety stanchions. They will be removed only after the aircraft director gives the signal. Tie-downs prevent the aircraft from moving in any direction. To do this, they must "oppose" each other and be as equally distributed on the aircraft as possible. Tie-down requirements are divided into three categories. Under normal conditions, these categories may be defined by the minimums discussed below.

(a) ***Initial (four-point) tie-down.*** This configuration is required for all aircraft before launch, when an aircraft is parked after recovery, or before the aircraft is moved. For armed OH-58D(I) aircraft, two chains (left rear and right front) are enough initially. This arrangement prevents the chain handler from having to work in front of loaded weapons pylons (rockets and a .50 caliber machine gun).

(b) ***Permanent (eight-point) tie-down.*** This configuration is required when the ship is not at flight quarters or when the aircraft is not expected to be respotted. The crew chief or PC applies permanent tie-downs.

(c) ***Heavy weather (twelve-point) tie-downs.*** This configuration is required when an increase in aircraft security is necessary because of high winds or sea state, ship's maneuvers, or for long periods of heavy maintenance.

NOTE: Different models of aircraft may require a different number of tie-downs. The aircraft handling officer may adjust the number of tie-downs needed in each of the above categories. He also may order an increase in the number of tie-downs because of the expected wind, sea state, or ship's maneuvers.

6-6. FUELING AND DEFUELING AIRCRAFT

a. The air officer is responsible to the ship's commanding officer for supervising and directing the receipt, stowage, and dispensing of aviation fuels. He also is responsible for the maintenance and security of the aviation fuels system and the enforcement of safety precautions. An effective aviation fuel quality control program is a vital part of aviation fuels system management.

b. The aviation fuels officer is responsible to the air officer for the efficient and safe operation of the aviation fuels system and for management of the aviation fuel quality control program. He also is responsible for ensuring strict compliance with all applicable technical directives concerning the inspection, maintenance, and operation of the aviation fuels system.

(1) ***Fueling and defueling procedures.***

(a) Normally, aircraft will be fueled as soon as possible after recovery. Each crew chief or PC will notify the aviation fuels petty officer or aviation fuels control talker in flight deck control if the fueling crew misses his aircraft. The crew chief or PC also will request that his aircraft be topped off as necessary after a maintenance run-up.

(b) Aircraft will be fueled according to the air plan. If a fuel load other the one shown in the air plan is desired, the squadron sends a request to air operations to have the air plan changed to the desired load. Requests for defueling for maintenance purposes are sent to the aircraft handler through the maintenance liaison officer.

(c) Fueling will be conducted in a way that causes the least interference with aircraft respotting. Before aircraft recovery, fueling crews will stand in or near their stations to break out hoses and start fueling aircraft. Before refueling begins, all aircraft and fuel hoses will be grounded properly. When fueling is completed, all ground wires will be removed. The aviation fuels officer will ensure that the appropriate smoking lamp condition is set before fueling or defueling begins.

(d) The crew chief or PC will ensure that the fuel load is correct and the filler caps are secure. The fuel control talker keeps the fuel status board in flight deck control. This board lists each aircraft on board and shows its exact fuel load.

(2) *Special safety precautions.*

- Aviation fuel will not be placed in open containers.
- Waste or rags soaked in aviation fuel will be disposed of properly and not left about the deck.
- Except for safety lights, no lights will be introduced into any compartment or space where aviation fuel or flammable fumes are present.
- Aviation fuel will not be transferred without notifying the engineering officer.
- Aviation fuel will not be discharged overboard without permission from the ship's commanding officer.
- If aviation fuel is spilled on the deck, the deck will be swabbed immediately and the incident reported to the aircraft handling officer.
- Lighted cigarettes or exposed flames of any kind will not be permitted near tanks, pipes, or containers of aviation fuel.
- Fuel will not be issued for any purpose other than fueling.
- Personnel will avoid breathing aviation fuel vapors over long periods.
- If aviation fuel contacts skin or clothing, personnel will wash with soap and water as soon as possible.
- To prevent eye injury, fuel handlers will wear protective goggles.
- All measures prescribed for quality control of the fuel being transferred will be complied with before the fuel is delivered.
- The smoking lamp will be out on the flight deck, hangar deck, and all-weather decks.
- Fire protection will be provided according to the NATOPS Aircraft Firefighting and Rescue Manual.
- All personnel must show caution and be alert for dangerous situations.
- Refueling will stop and the equipment secured when a fuel spill is noted. Refueling will not continue until the spill has stopped and the residue cleaned up.
- Only the members of the flight crew and ship's refueling crew who are needed to conduct fueling operations will be near the aircraft.
- Before the fueling nozzle is attached to the aircraft, a ground wire will be attached to the deck and then to the aircraft.

(3) *Hot refueling.* Aircraft that are equipped for pressure refueling may be hot refueled during training, operational, and combat situations. During hot refueling, the LSE or director will position himself where he can see the pilots, fueling station operator, and nozzlemen. He also will ensure that refueling personnel, equipment, chocks, and tie-downs are clear before he gives the launch or taxi signal to the pilot.

NOTE: Hot refueling on aircraft that require gravity refueling is not authorized.

(4) ***Hot refueling safety precautions.***

(a) Procedures for hot refueling will be according to the NAVSHIPs technical manual and applicable aircraft NATOPS or operator's manuals. The aircraft will be chocked and the initial tie-down applied. The tie-down crew will remain at the aircraft main mounts for rapid breakdown should an emergency launch be required.

(b) Personnel will move from one side of the aircraft to the other by way of the nose. Under no circumstances will personnel work close to a tail rotor.

(c) Passengers will leave the aircraft before hot refueling begins.

(5) ***Pressure refueling with the aircraft shut down.*** Pressure refueling with the aircraft shut down is the normal procedure. The aircraft will be completely shut down; only the PC, refueling party, and fire party need to remain on station. More information on pressure refueling is in the applicable aircraft operator's manual.

6-7. MEDICAL CASUALTIES ON THE FLIGHT DECK

Medical casualties brought on board by aircraft will be removed from the aircraft and handled according to the ship's casualty handling bill. The ship's medical department will be notified as soon as possible to allow medical personnel to meet incoming aircraft.

Section III. Weapons Handling

6-8. PROCEDURES

a. Airborne weapons handling evolutions introduce a degree of risk that requires careful planning and preparation. The necessity to train for and conduct combat operations requires that certain risks that cannot be avoided in the handling of explosive weapons be accepted. Commanding officers will continually weigh the need to perform each weapons evolution against each additional risk. Only those evolutions where the need outweighs the risk will be accepted.

b. Ordnance must be assembled and delivered to the flight deck in enough time and quantity to meet the air plan. Therefore, ordnance breakout and movement for assembly requires planning and close coordination between weapons personnel and the air department. Backloading requires the same coordination, but timing becomes less critical.

c. The weapons officer is responsible for the breakout of aviation ordnance that is specified in the air plan. Along with the aviation unit armament officer, he decides the quantities and types of aviation ordnance and the times that it will be delivered to the assembly area. By doing so, the weapons officer ensures that further movement proceeds in a safe and timely manner. He also is responsible for moving ordnance from the magazines to the assembly area and from the assembly area to the flight deck. Using crews from the embarked unit, the air officer is responsible for safe movement on the flight deck.

d. The ship's personnel are responsible for moving ordnance from the magazines to the assembly areas. Embarked personnel may help as necessary. The ship's personnel must assemble and move all weapons from the assembly area to the staging areas. Embarked personnel will move all weapons from the staging areas to the aircraft. Movements will be made using the safest and most direct route.

WARNING

Staging of ordnance in areas around the refueling-at-sea station is prohibited if refueling is being conducted while the ship is underway.

NOTE:

1. During ordnance handling evolutions above the second deck, personnel must comply with the AFFF system and mobile firefighting equipment information in NAVAIR 00-80R-14.

2. When required, electrical power may be applied during the aircraft loading and downloading procedure. It should, however, be held to a minimum consistent with operational requirements. Electrical power will not be applied to the armament or weapon release and control circuitry while weapons are being loaded or downloaded.

e. When aircraft are carrying weapons or ordnance, an EOD representative will be available at the flight-deck level during all launch and recovery operations.

f. During launch and recovery operations, the EOD representative and the designated air gunner and squadron ordnance representatives will be equipped with an SRC-22 (or equivalent) communications set.

6-9. HAZARDS OF ELECTROMAGNETIC RADIATION TO ORDNANCE AND SAFETY PRECAUTIONS

a. Modern radio and radar transmitting equipment produces high-intensity radio frequency fields. These fields can cause premature actuation of sensitive electroexplosive devices in the ordnance systems and biological injury to personnel working nearby. Sparks or arcs caused by high-intensity fields are a potential source of ignition for fuel-air mixtures. The most susceptible times are during assembly, disassembly, loading, unloading, or testing in electromagnetic fields.

b. The effect of premature operation of these devices varies with the function of the device initiated. The most likely effects are dudding, loss of reliability or, in the case of rockets and flares, ignition of the propellant or illuminant. In several electromagnetic radiation environments, a finite probability of warhead detonation exists. Therefore, the ship's electromagnetic environment must be controlled when HERO-susceptible ordnance is present or is being handled or unloaded.

c. Before embarkation, pilots, aircrews, and squadron ordnance personnel will familiarize themselves with the latest HERO conditions in NAVSEA OP 3565, NAVAIR 16-1-529, and the ship's HERO/EMCON bill. The technical manual on Radio Frequency Hazards to Ordnance, Personnel, and Fuel gives detailed operating procedures and precautions to include in the ship's EMCON bill.

d. Before starting operations that involve HERO-susceptible ordnance, the proper HERO condition must be set. The HERO condition will be readily displayed so that assembly, flight deck, and hangar deck ordnance personnel see the HERO condition at all times. The OOD will announce over the ship's public address system when the HERO-EMCON condition is set or canceled.

e. The ship's commanding officer will request a HERO survey, which is required by NAVSEA OP 3565 or NAVAIR l0l-529. When the survey is finished, a HERO/EMCON bill will be established.

6-10. WEAPONS MOVEMENT AND HANDLING

a. Airborne weapons outside of the designated magazines greatly increase the danger to the ship should a fire or explosion occur. The more weapons there are, the greater the risk. To reduce this risk, only the number of weapons needed to sustain operations will be transferred to the hangar or flight deck.

b. Weapons handlers must have enough time to load the aircraft safely. Therefore, airborne weapons must be placed where they are readily available. Staging areas for assembled or unassembled weapons will be restricted to areas that--

(1) Are supported by jettison ramps on the flight deck or an operable weapons elevator below the hangar deck or are within 50 feet of a jettison location on the hangar deck or sponson areas.

(2) Have at least two clear routes that are kept clear of obstructions in case of an emergency movement.

(3) Are covered by a water/deluge system or an operable sprinkler system or are protected by dedicated manned fire hoses.

(4) Are at least 10 feet from aircraft fueling stations and 20 feet from LOX facilities, converters, and carts.

(5) Are manned continually for rapid jettison by qualified and certified personnel.

c. The flight deck, hangar deck, and sponson that meet the criteria in **(1)** through **(5)** above are authorized staging areas. Handling or assembly areas outside of magazines, which may be supported by operable elevators rather than jettison facilities, also are authorized staging areas.

d. The maximum weapon density in staging areas will be limited to the quantities shown below.

(1) On the flight deck, weapons are limited to those required for the next two events. This includes the number of weapons that have been loaded or are in the process of being loaded or staged.

(2) On the hangar deck and sponsons, weapons are limited to those required for one event.

(3) In the handling areas, weapons are limited to those required for immediate strikeup or strikedown.

(4) In the assembly area, weapons are limited to those required to sustain operations.

(5) Flight deck and hangar deck staging of parachute flares is limited to those required for the next two events. Paraflares and dispensers will be downloaded, safed, and returned to stowage at daylight or when the operation that required their use is completed.

e. Staging areas will be used for ready service only, not for protracted stowage nor for extending the total weapons stowage capacity of the ship. All weapons in staging areas will be on mobile trucks or skids.

f. Before flight operations, all ordnance jettison ramps will be fully functional and exercised daily according to the applicable PMS. When ordnance is present, jettison ramps in the staging areas will be rigged and unobstructed at all times. All other ramps will be rigged when required by the weapons officer. Aircraft elevators supplement weapon elevators and expedite strikeup of weapons during heavy ordnance operations. Coordination and thorough preplanning between the weapons officer and aircraft handling officer is essential to meet load/plan requirements and ensure safety.

g. To ensure that safety standards are complied with during all aviation ordnance evolutions, a certified ordnance safety supervisor is assigned from the ship. These safety supervisors will be familiar with this manual and other applicable directives. They have the authority to stop any evolution if, in their judgment, safety is being jeopardized. An evolution that is stopped will not continue until the matter is resolved.

h. A properly equipped EOD and a ship's weapons representative will be readily available to give technical assistance to the aircraft handling officer on weapons and their disposal. The weapons flight deck safety petty officer and senior embarked squadron ordnance representative will maintain a status board. This status board confirms the type, quantity, and location of all weapons on the flight deck and/or aircraft. In addition, weapons cook-off data will be posted in plain view of the aircraft handling officer.

i. With the exception of actual loading evolutions, weapons on skids or trucks will be positioned fore and aft and manned continuously.

6-11. LHA CLASS WEAPONS HANDLING RESTRICTIONS

On LHA-1 class ships, the restrictions described below apply during the conduct of all live aviation ordnance procedures.

- Unless a bomb barrier is installed on the ship, the bomb assembly area at the top of the vehicle ramp will not be used for live ordnance.
- Only certified ordnance handlers, designated in writing by the commanding officer, will perform breakout, assembly, and staging of live aviation ordnance.

6-12. WEAPONS ASSEMBLY AND DISASSEMBLY

a. Because of the danger involved, the assembly and disassembly of aviation ordnance will be controlled closely. All unpacking, assembly, disassembly, loading, and unloading of weapons will be done according to NAVSEA OP 4, NAVSEA OP 3565/NAVAIR 16-1-529 and the appropriate checklists, SRCs, and technical manuals.

b. Ordnance will be assembled, disassembled, and loaded into launchers or magazines only by properly certified personnel. According to the OPNAVINST 8023.2-series, a safety supervisor will be present when ordnance is being assembled, loaded, unloaded, or disassembled. Normally, all assembly and disassembly will be done in the ordnance assembly area.

c. The assembly area will be kept HERO-safe whenever the ordnance is HERO-susceptible. If HERO-susceptible ordnance is moved outside the normal HERO-safe assembly area, the operations officer will ensure that the appropriate HERO condition has already been set. He also will ensure that the HERO condition has been set if assembly must be done in a HERO-unsafe area.

d. All rockets will be unpacked, assembled, loaded into, and unloaded from launchers in designated assembly areas only.

e. Ships will maintain technical manuals for each type of aviation weapon on board. All weapon systems maintenance will be done by squadron aviation ordnance technicians.

f. All personnel involved in unpacking, assembly, and disassembly will be appropriately certified.

6-13. WEAPONS LOADING AND DOWNLOADING

Guidance for weapons loading and downloading is provided in Appendix C.

a. The aircraft handling officer, ship's weapons officer, squadron ordnance personnel, and unit maintenance liaison officer must coordinate closely to comply with the weapons requirements in the air plan. As soon as possible, the unit ordnance officer advises the maintenance liaison officer of any special requirements that apply to loading selected aircraft.

b. The unit maintenance liaison officer ensures that the aircraft handling officer is informed of any special requirements, configuration, or status that may make certain aircraft unassignable for particular types of weapons loads.

c. After coordinating with squadron maintenance representatives, the aircraft handling officer designates the aircraft to be loaded. Adequate time must be allowed to make the required configurations and perform aircraft release and control system checks. Therefore, the aircraft handling officer provides ordnance personnel with the planned deck spot as early as possible.

d. Simultaneous fueling, loading and downloading of weapons and preloaded TERs, and the installation of fuses and arming wires is authorized.

WARNING

1. Oxygen servicing (other than replacing the converter at the aircraft), loading, and downloading are conducted as separate operations.

2. Loading forward-firing ordnance that requires a simultaneous and/or prior electrical connection is not authorized while fueling is in progress. No other electrical connections to weapons or removal or installation of impulse cartridges will be done while fueling is in progress. Fuel hoses will not be placed under weapons that are being loaded or downloaded.

3. Aircraft to be loaded with rockets and/or missiles should be positioned so that an accidental discharge will not endanger personnel, the ship, or other aircraft.

e. No-voltage checks will be made after normal rotor engagement when the electrical system is on aircraft power. The signal to start no-voltage checks will not be given until the flight deck OSS sees the copilot's hands and receives the copilot's acknowledgment. Any deviation from the above procedure must be according to the authorized weapons checklist.

NOTE: The flight deck is the preferred area to load and download aircraft.

f. At times, operational necessity may require that the added risk of fire caused by fuel and explosives in a confined area be accepted. In this case, the ship's commanding officer may authorize loading limited amounts of weapons on the hangar deck. Only aircraft that are scheduled for the next launch or ones that are in an alert status will be loaded on the hangar deck. The weapons to be loaded on these aircraft are restricted to the ones shown in Appendix C.

WARNING

1. While the engines are running, personnel will not approach the aircraft to perform weapon systems checks until the ordnance arming supervisor gives clearance. The ordnance arming supervisor will be in full view of the pilot and will have the pilot's attention.

2. Tube loading 2.75-inch rocket launchers is prohibited. (The Navy prefers to load rockets in the rocket pod while it is disconnected from the aircraft. Before the mission begins, the pod is winched onto the aircraft full of rockets. This is referred to as tube loading.)

NOTE:

1. The mechanical latching on aircraft racks and launchers will be completed before the engines on the aircraft are started.

2. The inert conventional weapons and captive air-launched missiles will be loaded and down-loaded and armed and dearmed the same way as live weapons.

6-14. ARMING

a. Weapons arming will be conducted in a designated arming area. When forward-firing weapons are involved and the NAVAIR weapons and stores loading checklists/SRCs require, the area ahead of the aircraft will be clear and kept clear until the launch is completed. Arming will be conducted only while the aircraft is at a complete stop and control of the aircraft given to an arming crew supervisor. All arming signals will be according to Appendix B of this manual.

b. The helicopter will be armed after the pilot signals that he is ready for takeoff and the tie-down chains and chocks have been removed. To provide the least hazard to arming crew members, the air gunner/ordnance officer or air boss will formalize the exit paths for each type of aircraft.

WARNING

Arming crews should use extreme caution when exiting armed aircraft to avoid injury from the aircraft rotors and intake/exhaust and the exhaust end of missile or rocket motors.

6-15. DEARMING

a. The designated aircraft dearming supervisor positions himself on the flight deck to ensure that the LSE/aircraft director and the dearming crew coordinate during recovery operations. The dearming supervisor indicates to the LSE/director those aircraft that require safing before they are moved or shut down.

b. Helicopters that land with hung weapons and/or forward-firing weapons will be dearmed before chain tie-downs are installed. They will be safed according to NAVAIR weapons/stores loading checklists/SRCs, and/or EOD emergency procedures. Aircraft safing signals will be according to Appendix B of this manual.

c. Aircraft landing with unexpended weapons will have those weapons safed according to NAVAIR weapons/stores loading checklists/SRCs. In all cases, they will be safed before any postflight checks or aircraft refueling begins. Appendix C lists weapons that are authorized for recovery.

WARNING

Aircraft downloading will not begin until the engines are shut down and the appropriate main and tail rotor tie-downs have been attached.

6-16. ABORT STRIKEDOWN

a. The flight deck is the preferred area for downloading weapons. If a loaded aircraft must be struck below, its weapons will be downloaded immediately after it reaches the hangar deck unless that aircraft--

• Is readily available for flight and scheduled for the next launch.

• Is in an alert condition.

• Requires maintenance or servicing that is allowed only on aircraft loaded with weapons.

b. In an abort/strikedown situation, the abort or after-landing procedures for the weapons prescribed in the NAVAIR weapons and stores loading checklists/SRCs will be done before the aircraft is moved to the hangar deck.

WARNING

Bomb rack ejector or jettison cartridges will be removed from all aircraft stations before or immediately after the aircraft is struck down to the hangar deck.

c. Certain weapons are specifically excluded from the provisions of this section. Appendix C shows a list of those weapons that may not be struck below while they are loaded on an aircraft.

6-17. MAINTENANCE ON LOADED AIRCRAFT

a. Maintenance will not be conducted on aircraft loaded with weapons. However, routine service and minor maintenance to ready the aircraft for the next launch may be conducted after applying the restrictions listed below are applied.

(1) Weapons will be safed to the maximum degree specified in the NAVAIR weapons/stores loading checklists/SRCs.

(2) If a WARNING placard and/or control stick cover is displayed prominently in the cockpit, the maintenance or servicing of loaded aircraft that requires application of electrical power is limited to--

• Refueling.

• Replacement and checkout of communications and navigation equipment.

• Replacement and checkout of engine performance and flight instruments.

• Engine run-up or rotor engagement check.

• Flight control and hydraulic system checks.

(3) Maintenance that requires electrical power to be applied to the armament or weapon release and control circuitry will not be done while weapons are loaded or are being loaded or downloaded.

(4) Aircraft that require extensive troubleshooting, engine removal, complete jacking, and so forth are not considered readily available for flight. These aircraft must be downloaded before maintenance begins.

b. Downloading includes removing all impulse cartridges from ejector racks or breeches. It also includes removing all rounds of ammunition from the feed chutes or feed mechanisms of internal guns.

APPENDIX A

AIRCRAFT HANDLING SIGNALS

This appendix provides a graphic portrayal of aircraft handling signals. It is divided into five sections and shows general signals, handling signals, refueling signals, aircraft elevator signals, and helicopter-specific signals respectively. Figure A-1 begins on the following page.

SIGNAL	DAY	NIGHT	REMARKS
①  AFFIRMATIVE (ALL CLEAR)	Hand raised, thumb up.	Same as day signal with addition of wands.	Conforms to ICAO signal.
②  NEGATIVE (NOT CLEAR)	Arm held out, hand below waist level, thumb turned downwards.	Same as day signal with addition of wands.	
③  PASS CONTROL	With both arms at should height, point in direction of person receiving control.	Same as day signal except point wands.	
④  THIS WAY	Arms above head in vertical position with palms facing inward.	Same as day signal with addition of wands.	Conforms to ICAO signal.

Figure A-1. General signals

SIGNAL	DAY	NIGHT	REMARKS
(5)  SLOW DOWN	Arms down with palms towards ground, then moved up and down several times.	Same as day signal with addition of wands.	Conforms to ICAO signal.
(6)  TURN TO LEFT	Point right arm downward, left arm is repeatedly moved upward - backward. Speed of arm movement indicates rate of turn.	Same as day signal with addition of wands.	Also used for spot turns for airborne aircraft. Conforms to ICAO signal.
(7)  TURN TO RIGHT	Point left arm downward, right hand repeatedly moved upward - backward. Speed of arm movement indicating rate of turn.	Same as day signal with addition of wands.	Also used for spot turns for airborne aircraft. Conforms to ICAO signal.
(8)  MOVE AHEAD	Arms extended from body and held horizontal to shoulders with hands up-raised and above eye level, palms facing backwards. Execute beckoning arm motion angled backward. Rapidity indicates speed desired of aircraft.	Same as day signal with addition of wands.	

Figure A-1. General signals(Continued)

SIGNAL	DAY	NIGHT	REMARKS
(9)  STOP	Arms crossed above the head, palms facing forward.	Same as day signal with addition of wands	
(10)  BRAKES	"ON"-Arms above head, open palms and fingers raised with palms toward aircraft, then fist closed. "OFF"- Reverse of above.	"ON"-Arms above head then wands crossed. "OFF"-Crossed wands, then uncrossed.	
(11)  MOVE BACK (ALSO USED TO PULL BACK AIRCRAFT USING ARRESTING WIRE)	Arms by sides, palms facing forward, swept forward and upward repeatedly to shoulder height.	Same as day signal with addition of wands.	Conforms to ICAO signal.
(12)  TURNS WHILE BACKING (TAIL TO LEFT)	Point right arm down and left arm brought from overhead, vertical position to horizontal position repeating left arm movement.	Same as day signal with addition of wands.	Conforms to ICAO signal.

Figure A-1. General signals(Continued)

SIGNAL	DAY	NIGHT	REMARKS
(13)  TURNS WHILE BACKING (TAIL TO RIGHT)	Point left arm down and right arm brought from overhead, vertical position to horizontal forward position, repeating right arm movement.	Same as day signal with addition of wands.	Conforms to ICAO signal.
(14)  CLEARANCE FOR PERSONNEL TO APPROACH AIRCRAFT	A beckoning motion with right hand at eye level.		
(15)  PERSONNEL APPROACHING THE AIRCRAFT	Left hand raised vertically overhead, palm towards aircraft. The other hand indicates to personnel concerned and gestures towards aircraft.	Same as day signal except point wands.	
(16)  INSERT CHOCKS	Arms down, fists closed, thumbs extended inward, swing arms from extended position inward.	Same as day signal with addition of wands.	Conforms to ICAO signal.

Figure A-1. General signals (continued)

SIGNAL	DAY	NIGHT	REMARKS
17  RAMP: OPEN/CLOSE (LSE)	One hand held in hold, the other finger and thumb extended but not touching, then bring fingers and thumb together several times. Pilot will respond with same signal.	Two wands used in same manner.	Ramp shall not come down until deck crew gives signal to pilot.
18  PASSENGERS READY FOR EMBARK/DISEMBARK (LSE)	Fingers used in walking motion.	Wands used in walking motion.	One handed signal from pilots.
19  REMOVE CHOCKS	Arms down, fist closed, thumbs extended outward, swing arms outward.	Same as day signal except point wands.	Conforms to ICAO signal.
20  INSTALL DOWN LOCKS/ UNDERCARRIAGE PINS	With arms above head, the right hand clasps left forearm and the left fist is clenched.	Similar to the day signal except the right wand is placed against left for arm. The wand in the left hand is held vertical.	

Figure A-1. General signals (continued)

SIGNAL	DAY	NIGHT	REMARKS
21  REMOVE DOWN LOCKS/ UNDERCARRIAGE PINS	With arms and hands in "INSTALL DOWN LOCKS" position, the right hand unclasps the left forearm.	Similar to the day signal except with the addition of wands.	
22  CONNECT GROUND ELECTRICAL POWER SUPPLY	Hands above head, left fist partially clenched, right hand moved in direction of left hand with first two fingers extended and inserted into circle made by fingers of the left hand.	Same as day signal with addition of wands.	
23  DISCONNECT GROUND ELECTRICAL POWER SUPPLY	Hands above head, left fist partially clenched, right hand moved away from left hand, withdrawing first two fingers from circle made by fingers of the left hand.	Same as day signal except point wands.	
24  START ENGINES	Left hand overhead with appropriate number of fingers extended, to indicate the number of the engine to be started, and circular motion of right hand at head level.	Similar to day signal except that the wand in the left hand will be flashed to indicate the engine to be started.	Conforms to ICAO signal.

Figure A-1. General signals (continued)

SIGNAL	DAY	NIGHT	REMARKS
25  SLOW DOWN ENGINES ON INDICATED SIDE	Arms down with palms toward ground, then either right or left arm waved up and down indicating that left or right side engines respectively should be slowed down.	Same as day signal with addition of wands.	Conforms to ICAO signal.
26  CUT ENGINES	Either arm and hand level with shoulder, hand moving across throat, palm downward. The hand is moved sideways with the arm remaining bent.	Same as day signal with addition of wands.	Conforms to ICAO signal.
27  READY AUXILIARY POWER UNIT START (PILOT)	Hand extended with three fingers extended.	Moves red flashlight behind the extended fingers	This signal will also be used prior to shutdown to restart APU.
28  LOCK TAIL/NOSE WHEEL	Hands together overhead, opened from the wrists in V, then closed suddenly.	Same as day signal with addition of wands.	

Figure A-1. General signals (continued)

SIGNAL	DAY	NIGHT	REMARKS
(29)  UNLOCK TAIL/NOSE WHEEL	Hands overhead, palms together, then hands opened from the wrists to form a V, wrists remaining together.	Same as day signal with addition of wands.	
(30)  FOLD WINGS/ HELICOPTER BLADES	Arms straight out at sides, then swept forward and hugged around shoulders.	Same as day signal with addition of wands.	
(31)  SPREAD WINGS/ HELICOPTER BLADES	Arms hugged around shoulders, then swept straight out to the side.	Same as day signal with addition of wands.	
(32)  LOCK WINGS/ HELICOPTER BLADES	Hit right elbow with palm of left hand.	Same as day signal with addition of wands.	

SIGNAL	DAY	NIGHT	REMARKS
33  OPEN WEAPON BAYS DOORS	Body bent forward at the waist, hands held with fingertips touching in front of body and elbows bent at approximately 45 degrees, then arms swing downward and outward.	Same as day signal with addition of wands	
34  CLOSE WEAPON BAYS DOORS	Body bent forward at the waist, and arms extended horizontally, then arms swing downward and in until fingertips touch in front of the body with elbows bent at approximately 45 degrees.	Same as day signal with addition of wands.	
35  TAKE OFF	Director conceals left hand and makes circular motion of right hand over head in horizontal planes ending in a throwing motion of arm towards direction of takeoff.	Same as day signal with addition of wands.	
36  ENGINE FIRE	Describes a large figure eight with one hand and points to the fire area with the other hand.	Same as day signal with addition of wands.	Signal is meant for information only. Pilot should be given a cut engine or continuous turn-up signal, as appropriate.

Figure A-1. General signals (continued)

SIGNAL	DAY	NIGHT	REMARKS
37  ENGAGE NOSEGEAR STEERING	Point to nose with index finger while indicating direction of turn with other index finger.	Same as day signal with addition of wands.	
38  DISENGAGE NOSEGEAR STEERING	Point to nose with index finger, lateral wave with open palm of other hand at shoulder height.	Same as day signal with addition of wands.	
39  LOWER WING FLAPS	Hands in front, palms together horizontally then opened from the wrist crocodile-mouth fashion.	Same as day signal with addition of wands.	
40  RAISE WING FLAPS	Hands in front horizontally, with palms open from the wrists, then suddenly closed.	Same as day signal with addition of wands.	

Figure A-1. General signals (continued)

SIGNAL	DAY	NIGHT	REMARKS
(41)  TILLER BAR/STEERING ARM IN PLACE	Hold nose with left hand, right hand moving horizontally at waist level. a. Affirmative signal immediately following means: MAN IS TENDING BAR. b. A negative signal immediately following means: NO ONE TENDING BAR.	Same as day signal with addition of wands.	
(42)  REMOVE CHOCKS AND/OR TIE-DOWNS (Pilot)	Swings arms apart, thumbs extended outward.	Using hand-held light or flashlight, gives on/off signals at one second intervals.	
(43)  REMOVE TIE-DOWNS (Director)	To tiedown crew: Makes wiping motion down left arm with right hand.	Same as day signal except point wands.	
(44)  INSERT CHOCKS AND/OR INSTALL TIE-DOWNS (Pilot)	Swings arms together, thumbs extended inward. In single-piloted aircraft, pilot may swing one arm alternately from each side, thumb extended inward.	Moves hand-held light or flashlight at eye level in a horizontal plane alternately inward from each side.	

Figure A-1. General signals (continued)

SIGNAL	DAY	NIGHT	REMARKS
45  INSTALL TIE-DOWNS (Director)	To tiedown crew: Rotates hands in a circle perpendicular to and in front of his body.	Same as day except with wands.	
46  TIE-DOWNS IN PLACE (Director)	Same signal as "install tiedown," followed by "thumbs up."	Same as day except with wands.	
47  ENGINE RUN-UP (Pilot)	Moves forefinger in a circular motion in view of director to indicate that he is ready to run up engine.	Makes circular motion with hand hold light.	Director responds with same signal (wand at night) to indicate "clear to run up."
48  HOT BRAKES	Makes rapid fanning motion with one hand in front of face and points to wheel with other hand.	Same as day except with wands.	

Figure A-1. General signals (continued)

SIGNAL	DAY	NIGHT	REMARKS
(49)  LIGHTS	Points to eyes with two fingers to signal "lights on."	Flashing wands.	When lights are already on, same signal is used to signal "lights off."
(50)  I HAVE COMMAND	Hold one hand open, motionless and high above head, with palm forward.	Same as day except with wand.	
(51)  START AIRCRAFT AUXILIARY POWER UNIT	Points to power unit exhaust with left index hand finger; moves right hand in horizontal circle, index and middle finger pointing downward.	Same as day signal except point wands.	
(52)  STOP AIRCRAFT AUXILIARY POWER UNIT	Makes "throat cutting" action with left hand; moves right hand in horizontal circle, index and middle fingers pointing downward.	Same as day except with wands.	

Figure A-1. General signals (continued)

SIGNAL	DAY	NIGHT	REMARKS
53  FUEL DISCHARGE DURING START	Left arm raised above shoulder with number of fingers extended to indicate affected engine; right hand describes a pendulum motion between waist and knees.	Similar to day signal except that wand in left hand will be flashed to indicate the number of the affected engine.	Signal is for information only; pilot should be given cut engine or continuous turnup signal, as appropriate.
54  ARM WATER INJECTION (AV-8)	Give FINAL TURNUP signal. Wait 2 or 3 seconds while pilot turns up to military rated thrust and checks instruments. Then, hold open hand toward pilot, fingers extended vertically.	Same except hold GREEN wand vertically and move up and down.	Day - Pilot acknowledges by salute. Night - Pilot acknowledges by turning on light to steady dim.
55  ENGINE THRUST CHECK (AV-8)	Extend arm overhead, forefinger pointing up. Hesitate, then rotate hand rapidly in a horizontal circle.	Hold RED and GREEN wands at chest level, rotating the green wand in a horizontal circle.	Signal is optional, given at request of pilot. Also can be used for deck launch.
56  FINAL TURN UP	Extend arm overhead, index and middle finger pointing up. Hesitate, then rotate hand rapidly in a horizontal circle.	Rotate GREEN wand in a horizontal circle at chest level. Hold RED wand behind back.	The pilot will apply or maintain full power, check instruments, get set and: Day - Turn head slightly toward launching officer, execute a hand salute, and position head against headrest. Night- Turn on only his running lights to steady and keep them on until clear of ship.

Figure A-1. General signals (continued)

SIGNAL	DAY	NIGHT	REMARKS
57  VTO (AV-8)	Arms extended horizontally sideways beckoning upwards with palms turned up.	Same as day signal with addition of wands	
58  ROTATE NOZZLES	Hold forearm vertically while nodding clenched fist followed by extend-ed number of fingers corresponding to each 10 degrees of rotation. Clenched fist for nozzles aft.	Same as day: Illuminate fingers with wands.	
59  TRIM CHECKS	Hold forearm horizontally while rubbing fingers together. Point down to get the stabilator loading edge down (increase positive trim); go up to get the leading edge up (decrease positive trim).	Same as day: Illuminate fingers with wands.	
60  SEAT PINS	Hold hand with index finger extended and point towards head and withdraw finger.	Same as day: Illuminate fingers with wands.	

Figure A-1. General signals (continued)

SIGNAL	DAY	NIGHT	REMARKS
61  EJECTION SEAT ARMED	From a vertical position, extend arm horizontally	Same as day: Illuminate fingers with wand.	
62  LAUNCH	Extend arm overhead. Ensure dock is clear forward. Sweep up-raised hand downward in the direction of launch, touching the deck and returning hand to the horizontal in the direction of launch.	Raise green wand vertically overhead. Wait 2 or 3 seconds to ensure that deck is clear forward. Sweep wand in a wide arc, ending by pointing in the direction of launch touching the deck and returning wand to the horizontal in the direction of launch.	
63  SUSPEND	Cross arms high over head indicating the launch is off.	Hold red wand overhead indicating the launch is off. Green wand is turned off.	

Figure A-1. General signals (continued)

SIGNAL	DAY	NIGHT	REMARKS
64  THROTTLE	Hold one fist at waist level, thumb extended up. Grasp thumb with other hand and rock as if adjusting throttle.	Hold red wand horizontally across chest. Raise and lower wand horizontally. Green wand is off.	
65  NOZZLES BREAKING STOP	Signalman gives rotate nozzles signal and then extends arms forward, parallel to deck, palms down, and sweeps the forearms down.		

Figure A-1. General signals (continued)

LEGEND:

1. Affirmative (All Clear)

2. Negative (Not Clear)

3. Pass Control

4. This Way

5. Slow Down

6. Turn to Left

8. Turn to Right

9. Stop

10. Brakes

11. Move Back (Also used to pull back aircraft using arresting wire)

12. Turns While Backing (Tail to Left)

13. Turns While Backing (Tail to Right)

14. Clearance for Personnel to Approach Aircraft

15. Personnel Approaching the Aircraft

16. Insert Chocks

17. Ramp: Open/Close (LSE)

18. Passengers Ready for Embark/Disembark (LSE)

19. Remove Chocks

20. Install Down Locks/Undercarriage Pins

21. Remove Down Locks/Undercarriage Pins

22. Connect Ground Electrical Power Supply

23. Disconnect Ground Electrical Power Supply

24. Start Engines

25. Slow Down Engines on Indicated Side

26. Cut Engines

27. Ready Auxiliary Power Unit Start (Pilot)

28. Lock Tail/Nose Wheel

29. Unlock Tail/Nose Wheel

30. Fold Wings/Helicopter Blades

31. Spread Wings/Helicopter Blades

32. Lock Wings/Helicopter Blades

33. Open Weapon Bays Doors

34. Close Weapon Bays Doors

35. Take Off

36. Engine Fire

37. Engage Nosegear Steering

38. Disengage Nosegear Steering

39. Lower Wing Flaps

40. Raise Wing Flaps

41. Tiller Bar/Steering Arm in Place

42. Remove Chocks and/or Tie-Downs (Pilot)

43. Remove Tie-Downs (Director)

44. Insert Chocks and/or Install Tie-Downs (Pilot)

45. Install Tie-Downs (Director)
46. Tie-Downs in Place (Director)
47. Engine Run-Up (Pilot)
48. Hot Brakes
49. Lights
50. I Have Command
51. Start Aircraft Auxiliary Power Unit
52. Stop Aircraft Auxiliary Power Unit
53. Fuel Discharge During Start
54. Arm Water Injection (AV-8)
55. Engine Thrust Check (AV-8)
56. Final Turn Up
57. VTO (AV-8)
58. Rotate Nozzles
59. Trim Checks
60. Seat Pins
61. Ejection Seat Armed
62. Launch
63. Suspend
64. Throttle
65. Nozzles Breaking; Stop

Figure A-1. General signals (continued)

SIGNAL	DAY	NIGHT	REMARKS
(1)  TILLER BAR IN PLACE OR ENGAGE NOSE WHEEL STEERING (Director)	Touch end of nose with forefinger. Then, give "thumbs up" signal with same hand.	Touch end of nose with wand. Then, give "up" signal with same wand.	
(2)  TILLER BAR REMOVED OR DISENGAGE NOSE WHEEL STEERING (Director)	Touch end of nose with forefinger. Then, sweep arm downward in direction of aircraft movement.	Touch end of nose with wand. Then, sweep wand downward in direction of aircraft movement.	
(3)  WING RIDER (Director)	Position forearms flat against each other in front of and perpendicular to body.	Same, holding wands.	
(4)  INSTALL TIE-DOWNS (Director)	Rotate hands in a vertical circle in front of body.	Same as day except with AMBER wands.	

Figure A-2. Handling signals

SIGNAL	DAY	NIGHT	REMARKS
(5)  TIE-DOWNS IN PLACE (Director)	Same as "INSTALL TIEDOWNS," adding a "thumbs up" signal.		
(6)  REVERSE THRUST TAXI (Director)	Palms facing aircraft at eye level with a push back motion. For turns, the director points in direction tail is to move.	Same, holding wands.	

LEGEND:

1. Tiller Bar in Place/Engage Nose Wheel Steering (Director)
2. Tiller Bar Removed/Disengage Nose Wheel Steering (Director)
3. Wing Rider (Director)
4. Install Tie-Downs (Director)
5. Tie-Downs in Place (Director)
6. Reverse Thrust Taxi (Director)

Figure A-2. Handling signals (continued)

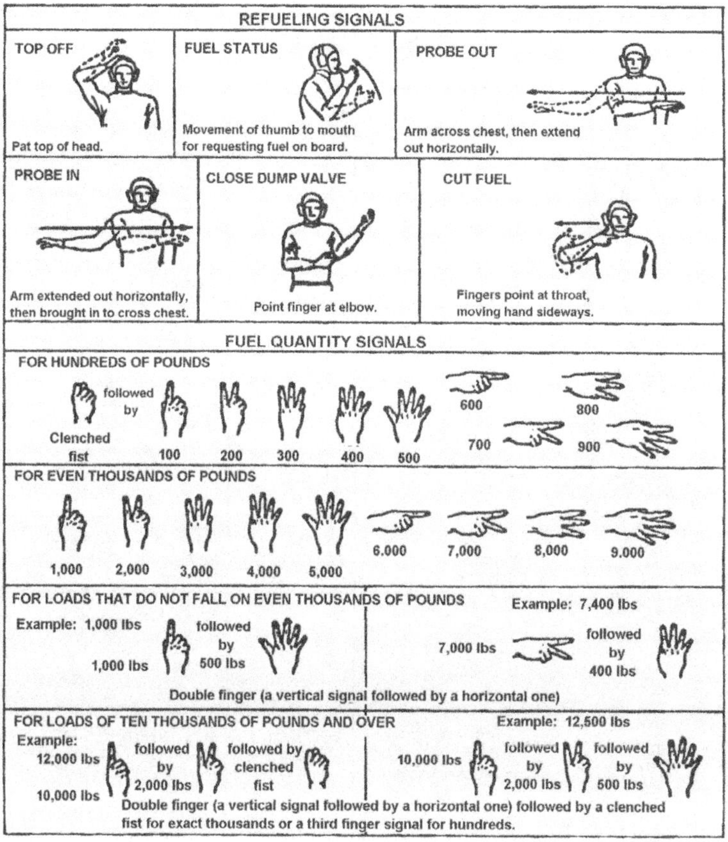

Figure A-3. Refueling signals

SIGNAL	FROM	TO	EXECUTION
① RAISE SAFETY STANCHION  	Elevator safety petty officer/director	Elevator operator	Day: Raise both index fingers extended upward chest level, close together, near body. Night: Raise both wands pointed upward at shoulder level, close together, near body.
② LOWER SAFETY STANCHION  	Elevator safety petty officer/director	Elevator operator	Day: Lower both index fingers, extended downward, chest level, close together, near body. Night: Lower both wands pointed downward at waist level, close together, near body.
③ RAISE ELEVATOR  	Elevator safety petty officer/director	Elevator operator	Day: Fully extend both arms with index finger pointing upward. Night: Fully extend both arms with wands pointing upward.
④ LOWER ELEVATOR  	Elevator safety petty officer/director	Elevator operator	Day: Fully extend both arms with index finger pointing downward. Night: Fully extend both arms with wands pointing downward.

Figure A-4. Aircraft elevator signals

LEGEND:

1. Raise Safety Stanchion
2. Lower Safety Stanchion
3. Raise Elevator
4. Lower Elevator

Figure A-4. Aircraft elevator signals (continued)

SIGNAL	DAY	NIGHT	REMARKS
(1)  LANDING DIRECTION	Marshaler stands with arms raised vertically above head and facing toward the point where the aircraft is to land. The arms are lowered repeatedly from a vertical to a horizontal position, stopping finally in the horizontal position.	Same as day signal with addition of wand.	
(2)  MOVE UPWARD	Arms extended horizontally sideways beckoning upwards, with palms turned up. Speed of movement indicates rate of ascent.	Same as day signal with addition of wand.	Conforms to ICAO signal.
(3)  HOVER	Arms extended horizontally sideways, palms downward.	Same as day signal with addition of wand.	Conforms to ICAO signal.
(4)  MOVE DOWNWARD	Arms extended horizontally sideways beckoning downwards, with palms turned down. Speed of movement indicates	Same as day signal with addition of wands.	Conforms to ICAO signal.

	rate of descent.		

Figure A-5. Helicopter-specific signals (continued)

SIGNAL	DAY	NIGHT	REMARKS
5  MOVE TO LEFT	Right arm extended horizontally sideways in direction of movement and other arm swung over the head in same direction, in a repeating movement.	Same as day signal with addition of wands.	
6  MOVE TO RIGHT	Left arm extended horizontally sideways in direction of movement and other arm swung over the head in the same direction, in a repeating movement.	Same as day signal with addition of wands.	
7  LOWER WHEELS	When aircraft approaches director with landing gear retracted, marshaler gives signal by side view of a cranking circular motion of the hands.	Same as day signal with addition of wands.	
8  WAVE-OFF	Waving of arms over the head.	Same as day signal with addition of wands.	

Figure A-5. Helicopter-specific signals (continued)

SIGNAL	DAY	NIGHT	REMARKS
9  LAND	Arms crossed and extended downwards in front of the body.	Same as day signal with addition of wands	Conforms to ICAO signal.
10  DROOP STOPS OUT	When rotor starts to "run down" marshaler stands with both hands raised above head, fists closed, thumbs pointing out.	Same as day signal with addition of wands	
11  DROOP STOPS IN	When droop stops go in, marshaler turns thumbs inward.	Same as day signal with addition of wands.	
12  REMOVE BLADE TIE-DOWNS	Left hand above head, right hand pointing to individual boots for removal.	Same as day signal with addition of wands.	

Figure A-5. Helicopter-specific signals (continued)

SIGNAL	DAY	NIGHT	REMARKS
(13)  ENGAGE ROTORS	Circular motion in horizontal plans with right hand above head.	Same as day signal with addition of wands.	
(14)  HOOK UP LOAD	Rope climbing motion with hands.	Same as day signal with addition of wands.	
(15)  RELEASE LOAD	Left arm extended forward horizontally, fist clenched, right hand making horizontal slicing movement below the left fist, palm downward.	Same as day signal with addition of wands.	
(16)  LOAD HAS NOT BEEN RELEASED	Bond left arm horizontally across chest with fist clenched, palm downward; open right hand pointed up vertically to center of left fist.	Same as day signal with addition of wands.	

Figure A-5. Helicopter-specific signals (continued)

SIGNAL	DAY	NIGHT	REMARKS
17  HOIST UP	Left arm horizontal in front of body, fist clenched, right hand with palm turned upwards making upward motion.	Same as day signal with addition of wands.	
18  HOIST DOWN	Left arm horizontal in front of body, fist clenched, right hand with palm turned down making downward motion.	Same as day signal with addition of wands.	
19  CUT CABLE	A signal similar to "RELEASE LOAD" except that the right hand has the palm downward and not clenched. Rapid repetition of right hand movement indicates urgency.	Same as day signal with addition of wands.	
20  SPREAD PYLON	Bend elbow across chest, palm downward. Extend arm outward to horizontal position, keeping palm open and facing down.	Same as day signal with addition of wands.	

Figure A-5. Helicopter-specific signals (continued)

SIGNAL	DAY	NIGHT	REMARKS
(21)  FOLD PYLON	Extend right arm horizontally, palm downward. Bend arm keeping palm down.	Same as day signal with addition of wands.	
(22)  I DESIRE HIFR/FUEL	Helo crew member brings thumb to mouth as if drinking from glass.	Same except use red lens flashlight.	
(23)  COMMENCE FUELING	Helo crew member makes circular motion with right hands.	Helo crew member makes circular motion with red lens flashlight.	
(24) 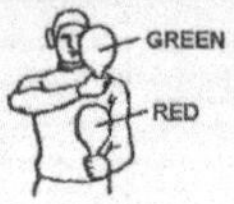 AM PUMPING FUEL	Ship's fuel crew member holds green device vertically over red device.	Ship's fuel crew member holds green wand vertically over red wand.	

Figure A-5. Helicopter-specific signals (continued)

SIGNAL	DAY	NIGHT	REMARKS
25  CEASE FUELING	Helo crew member makes horizontal cutting motion of right hand across throat.	Helo crew member makes horizontal motion of red lens flashlight.	
26 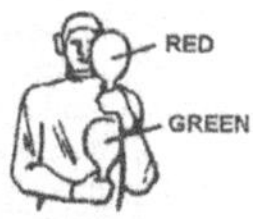 HAVE CEASED PUMPING FUEL	Ship's fuel crew member holds red device over green device.	Ship's fuel crew member holds red wand vertically over green wand.	
27  DESIRE TO MOVE OVER DECK AND RETURN HOSE	Crew member makes vertical motion of hand.	Helo crew member makes vertical motion of red lens flashlight.	
28  EXECUTE EMERGENCY BREAKAWAY	LSE/Director makes wave off signal.	LSE/Director makes wave off signal with wands.	Signal is mandatory.

Figure A-5. Helicopter-specific signals (continued)

SIGNAL	DAY	NIGHT	REMARKS
29  READY TO START ENGINE (Pilot)	Moves hand in a circle perpendicular to the deck; follows with a "thumbs up" signal. Signify by number of fingers engine to be started.	Turns on flashlight or movable light and moves it in a circle perpendicular to the deck.	
30  READY TO ENGAGE ROTORS (Pilot)	Moves hand in horizontal circle at eye level, index finger extended. Aircraft lights flashing bright.	Same as day except holds red light in hand. Aircraft lights flashing dim.	At night, aircraft lights should be on "flashing dim" until aircraft is declared up and ready for takeoff by the pilot.
31  READY TO ENGAGE ROTORS (LSE)	FACES FLY CONTROL: Holds left fist above head; gives circular motion of right hand above head, index finger extended.	Rotates one wand at chest level; holds other wand above head.	The air officer shall signal authority to engage rotors by illuminating a yellow rotating beacon.
32  READY FOR TAKEOFF (Pilot)	Gives "thumbs up" signal at eye level. Aircraft lights steady bright.	Places running and formation lights on "steady dim." May give "thumbs up" signal by turning on flashlight or other moveable lights and moving it up and down.	

Figure A-5. Helicopter-specific signals (continued)

SIGNAL	DAY	NIGHT	REMARKS
33  READY FOR TAKEOFF (LSE)	FACES FLY CONTROL. Holds right thumb up at eye level; holds left fist at eye level.	Signal not required. Pilot's "steady dim" indicates readiness to Fly Control.	The air officer shall signal authority for launch of helicopters by illuminating a green rotating beacon in addition to the rotating yellow beacon.
34  REMOVE TIE-DOWNS (LSE)	To tiedown crew: Makes wiping motion down left arm with right hand.	Same as day except holds amber wands.	
35 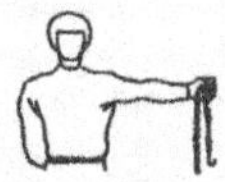 TIE-DOWNS REMOVED (Deck Crew)	Stands in full view of pilot and LSE and holds tiedown extended to sides with the chock on the deck in front.	Same as day except illuminates tiedown with amber flashlight.	
36  INSTALL TIE-DOWNS (LSE)	To tiedown crew: Rotates hands in a circle perpendicular to and in front of his body.	Same as day except with amber wands.	Give "Hold" signal as soon as first tiedown is attached.

Figure A-5. Helicopter-specific signals (continued)

SIGNAL	DAY	NIGHT	REMARKS
37  DISENGAGE ROTORS (LSE)	Holds left fist above head; makes throat cutting action with right hand.	Same as day except with amber wands.	Give "Hold" signal as soon as first tiedown is attached.
38  HOOK NOT DOWN/UP	Arms extended, make short up and down chopping action, alternating hands.	Same as day except with amber wands.	
39  HOLD POSITION	Makes clenched fist at eye level.	Hold crossed wands (any color) overhead.	Signal is mandatory.
40  RETURN TO FLIGHT/ GROUND IDLE (Pilot)	Makes fanning motion in front of face. May request with fingers desired engine.	Same signal with flashlight.	Turn down from 100% to flight/ ground idle prior to shutdown.

Figure A-5. Helicopter-specific signals (continued)

SIGNAL	DAY	NIGHT	REMARKS
41  CLEARED TO FLIGHT/ GROUND IDLE (LSE)	Make fanning motion in front of face. May request with one finger or two for desired engine.	Same signal with wand.	To indicate rotor is slowing down.

LEGEND:

1. Landing Direction
2. Move Upward
3. Hover
4. Move Downward
5. Move to Left
6. Move to Right
7. Lower Wheels
8. Wave-Off
9. Land
10. Droop Stops Out
11. Droop Stops In
12. Remove Blade Tie-Downs
13. Engage Rotors
14. Hook Up Load
15. Release Load
16. Load Has Not Been Released
17. Hoist Up
18. Hoist Down
19. Cut Cable
20. Spread Pylon

21. Fold Pylon
22. I Desire HIFR/Fuel
23. Commence Fueling
24. Am Pumping Fuel
25. Cease Fueling
26. Have Ceased Pumping Fuel
27. Desire to Move Over Deck and Return Hose
28. Execute Emergency Breakaway
29. Ready to Start Engine (Pilot)
30. Ready to Engage Rotors (Pilot)
31. Ready to Engage Rotors (LSE)
32. Ready for Takeoff (Pilot)
33. Ready for Takeoff (LSE)
34. Remove Tie-Downs (LSE)
35. Tie-Downs Removed (Deck Crew)
36. Install Tie-Downs (LSE)
37. Disengage Rotors (LSE)
38. Hook Not Down/Up
39. Hold Position
40. Return to Flight/Ground Idle (Pilot)
41. Cleared to Flight/Ground Idle (LSE)

Figure A-5. Helicopter-specific signals (continued)

APPENDIX B

AIRCRAFT ARMING AND SAFING SIGNALS

This appendix shows a graphic portrayal of aircraft arming and safing signals (Figure B-1). Signals from the arming and safing supervisor should be held in the position shown until the particular action (arming, safing, stray voltage) is completed. Figure B-2 shows the standardization of wands for arming and dearming personnel. Figure B-1 begins on the following page.

SIGNAL		MEANING	RESPONSE
DAY	NIGHT		
ARMING SUPERVISOR: Hands over head with finger tips touching. 	Red banded wands overhead with tips touching.	Pilot/Copilot/NFO: check all armament switches OFF or SAFE.	Pilot/Copilot/NFO: Raise both hands into view of arming super visor after checking switch positions. (Hands remain in view during check and hook up.)
ARMING SUPERVISOR: One hand over head; point to arming crew members with other hand. 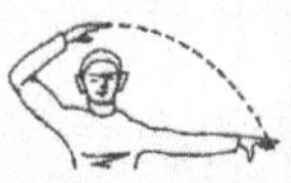	Same as day but with red banded wands.	Arming crew: Perform stray voltage checks.	Arming crew: Give thumbs up to arming supervisor if no stray voltage exits. Thumbs down indicates stray voltage problems. Night: Vertical sweep with flashlight indicates no stray voltage. Horizontal sweep indicates stray voltage.
ARMING SUPERVISOR: Raise fist, extended upward to meet horizontal palm of other hand. 	Form a tee with red banded wands.	Arming crew: Arm weapons (as applicable).	Arming crew: Give arming supervisor thumbs up when arming completed and clear immediate area. Thumbs down if malfunction exists. Night: Vertical sweep with flashlight indicates arming completed. Horizontal sweep indicates malfunction.

Figure B-1. Aircraft arming and safing signals

SIGNAL		MEANING	RESPONSE
DAY	NIGHT		
ARMING SUPERVISOR: Raise both hands with fingers pointing to sound attenuators. 	Same as day. Tips of red banded wands touching sound attenuators.	Arming crew: Perform missile check.	Pilot: Give arming supervisor thumbs up if tone is heard. Thumbs down if no tone. Night: Same as step 3 above.
ARMING SUPERVISOR: Insert finger of one hand into clenched fist of other hand and give extracting motion. 	Touch tips of red banded wands in front of body. Then move one wand laterally in a sweeping motion.	Arming crew: Remove bomb rack/pylon safety pins.	Arming crew: Shows pins to arming supervisor and clear immediate area. Night: Same as step 3 above.
ARMING SUPERVISOR: Give pilot-- (a) Thumbs up.  (b) Thumbs down. 	(a) Vertical sweep with red banded wand. (b) Horizontal sweep with red banded wand.	Pilot: (a) Aircraft armed and all personnel and equipment clear. (b) Aircraft down for weapons.	Pilot: (a) Acknowledge with similar signal. (b) Acknowledge with similar signal.

Figure B-1. Aircraft arming and safing signals (continued)

SIGNAL		MEANING	RESPONSE
DAY	NIGHT		
SAFING SUPERVISOR: **Hands over head with finger tips touching.** 	Red banded wands over head with tips touching.	Pilot/Copilot/NFO: Check all armament switches OFF or SAFE.	Pilot/Copilot/NFO: Raise both hands into view of safing supervisor after checking switch position. (Hands remain in view during safing.)
SAFING SUPERVISOR: **One hand over head, point to safing crew member with other hand.** 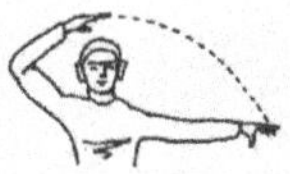	Same as day but with red banded wands.	Safing Crew: Safe weapons (as applicable).	Safing Crew: After safing give safing supervisor thumbs up and move clear of aircraft. Night: Vertical sweep with flashlight when safing is complete.
SAFING SUPERVISOR: **Give pilot "thumbs up."** 	Vertical sweep with red banded wand.	Pilot: Aircraft is safed and crew and equipment are clear.	Pilot: Acknowledge with similar signal.

Figure B-1. Aircraft arming and safing signals (continued)

PERSONNEL	COLOR	NO.	TYPE
Ordnance Arming/Safety Supervisor	Red	2	Standard Banded ***
Ordnance Arming Crew	Red	1	Stubby Banded **

* Standard and stubby cone shape.
Standard denotes full length cones; stubby is a modified cone providing 3 inches of lighted cone. Any suitable battery and switch housing is authorized if cone is brightly lighted. All signal wands/flashlights shall be equipped with heat-shrinkable sleeving to prevent possible cone separation.

** One 3/4 inch band on the cone (plastic electricians tape is recommended).

*** Two 3/4 inch bands spaced equidistant on the cone (plastic electricians tape is recommended).

Figure B-2. Standardization of wands for arming and dearming personnel

APPENDIX C

WEAPONS LOADING, STRIKEDOWN, DOWNLOADING, AND RECOVERY GUIDE

Safe handling of ordnance and explosive devices requires the attention of all echelons of command. Qualification and certification of aviation personnel in the safe, efficient handling of ordnance and explosive devices should be structured around existing training programs.

C-1. PERSONNEL

a. *Qualification Procedures.* Qualification of personnel will be as team members (TM), individuals (I), team leaders (TL), quality assurance personnel (QA), instructors (IN), and safety observers (SO).

b. *Certification.* The Navy will certify each person who is qualified and recommended for certification.

(1)*Team member.* All personnel whose duties require the handling, packaging, unpacking, assembling or disassembling, fuzing, loading, downloading, arming, or dearming of ordnance and explosive devices will be qualified and certified as team members. This level indicates an in-training status and applies to personnel who must be supervised while they perform their duties.

(2)*Individual.* Personnel whose duties require that they individually inspect (including acting as safety observers), prepare, adjust, arm, or dearm ordnance and explosive devices will be qualified and certified for such tasks. Personnel conducting magazine inspections; maintenance on aircraft, safety, and survival equipment; or performing any function that involves ordnance and explosive devices will be included in this program. Supervisors of ordnance and explosive device operations will be certified individually for operations that they may supervise or observe.

(3) *Team leader.* Personnel who have been qualified previously and certified to the I level and whose duties require that they direct and supervise others in safe and reliable operations may be designated TL.

(4) *Quality assurance.* This qualification and certification will be certified to the I or TL level. These personnel must have detailed knowledge of applicable ordnance and explosive devices or systems inspection criteria to determine that the device or system will function properly. Personnel also must be able to determine that the necessary assembly or installation procedures have been completed using applicable directives. Personnel who are quality assurance representatives or collateral duty quality assurance representatives and perform functions involving explosive devices also will be qualified and certified, as a minimum, to the QA level and as SOs as outlined below.

(5) *Safety observer.* The qualification and certification standards of the SO will ensure that the member knows the applicable safety procedures, equipment, and devices under his observation well enough to recognize and react to violations.

(6) *Instructor.* To obtain an IN qualification and certification, personnel will be qualified and certified as I or TL and have developed the necessary skills to instruct others using a command-approved course of instruction.

c. ***Revocation and Duration of Qualification and Certification.*** The commander of the embarked Army unit will determine the expiration date and cause for revocation.

C-2. EXPLOSIVE DEVICES

Each type of explosive device is considered a separate family. The following list of types of explosive devices is representative but not inclusive:

- High-explosive bombs and components.
- CBUs.
- Special-purpose bombs (practice bombs with marker charges, leaflet chaff).
- Pyrotechnics.
- Chemical ammunition.
- Demolition explosives and material.
- Mines and components.
- Cartridges and cartridge-actuated devices.
- Rocket warheads and components.
- Small arms and landing force ammunition.
- Aircraft gun ammunition.
- Air-launched guided missiles and components.
- Targets and components.
- Guided weapons.

C-3. EXPLOSIVE OPERATIONS

Each type of explosive operation is considered a separate family. The following list of explosive operations is representative but not inclusive.

a. ***Ashore Operations.***

- Receipt, segregation, storage, and issue functions.
- Aircraft arming and dearming.

b. ***Afloat Operations.***

- Aircraft release and control system checks.
- Aircraft loading and downloading.
- Aircraft arming and dearming.
- Ordnance and explosive device handling and transporting.

- Ordnance and explosive device unpackaging and packaging.
- Ordnance and explosive device inspection, assembly, or disassembly.
- Aircraft gun handling, loading, or jam clearing.
- Storage.

C-4. RECORD OF CERTIFICATION

Certifications will be documented using the appropriate Navy forms. Figure C-1 on the next page shows a sample format for a record of certification form. These samples may be included in the aviation unit SOP.

NAME:	RANK: SSN:	
CERTIFICATION LEVELS	**WORK TASK LEVEL**	
I: INDIVIDUAL TM: TEAM MEMBER TL: TEAM LEADER QA: QUALITY ASSURANCE SO: SAFETY OBSERVER IN: INSTRUCTOR	1. STORAGE/STOWAGE 2. HANDLING 3. ASSEMBLY/DISASSEMBLY 4. UPLOAD/DOWNLOAD 5. ARM/DEARM	6. TRANSPORTING 7. MAGAZINE INSPECTION 8. INSTALL/REMOVE 9. AIRCRAFT RELEASE

EXPLOSIVE DEVICE/ WEAPON	WORK TASK LEVEL	CERTIFICATION LEVEL	INSTRUCTOR SIGNATURE	DATE
CANNON/MACHINE GUN				
2.75" ROCKETS				
ATAS				
HELLFIRE MISSILE				
EXPLOSIVE CARTRIDGES				

RECERTIFICATION DATE:	EVALUATOR:
RECERTIFICATION DATE:	EVALUATOR:

1. Certification will last for 12 months.
2. Detachment OIC/NCOIC or Armament PSG are designated as trainers.

3. Company commanders are designated as Board Chairmen.

______________________________ is designated as shipboard-certified on ________________ to perform at the assigned work task level and work task codes as indicated.

______________________ BOARD CHAIRMAN/COMMANDER

Figure C-1. Sample format for a record of certification form

TASK	TASK TITLE	INST INITIALS	TASK	TASK TITLE	INST INITIALS
001	REMOVE/INSTALL WEAPONS		006	ASSEMBLE/DISASSEMBLE WPNS	
A	CANNON		A	CANNON	
B	2.75" ROCKETS		B	AMMO CAN/CHUTE	
C	HELLFIRE		007	HANGAR DECK SPOTTING	
D	ATAS		A	HANGAR	
E	EXPLOSIVE CARTS		(1)	STARBOARD	
002	UPLOAD/DOWNLOAD WPNS		(2)	PORT	
A	CANNON		B	DECK	
B	2.75" ROCKETS		(1)	SINGLE SPOT	
C	HELLFIRE		(2)	DUAL SPOT	
D	ATAS		(a)	DDG (RAST/NON-RAST)	
003	ARM/DEARM WEAPONS		(b)	FFS(RAST/NON-RAST)	
A	CANNON		C	HANGAR SWAP	
B	2.75" ROCKETS		008	CHAIN/UNCHAIN/ CHOCKS	
C	HELLFIRE		A	AIRCRAFT MOORING POINTS	
D	ATAS		B	DECK CLEATS	
E	AIRCRAFT RELEASE		C	VISUAL/LIGHT SIGNALS	
004	SHIPBOARD ARMAMENT		D	DECK PATH TO AVOID WPNS	

A	AMMO HANDLING		009	AIRCRAFT POSITION, MOVEMENT	
B	AMMO STORAGE		A	COMMO (LSE, SHIP, AIRDET)	
C	CROSSDECK AMMO		B	DET COMMUNICATIONS	
D	MAGAZINE INSPECTION		C	MOVEMENT/POSITIONS	
005	SHIPBOARD SPECIFIC		D	COMMANDS	
A	EMERGENCY 02 DEVICE		(1)	CHAINS OFF	
B	LPU		(2)	WHEELS UP	
C	O.B.A.		(3)	TAIL UP	
D	BERTHING EGRESS		(4)	PUSH	
E	EMERGENCIES		(5)	TAIL LEFT/RIGHT	
(1)	MAN OVERBOARD		(6)	LOWER/DROP IT	
(2)	FIRE		(7)	CHAINS ON	
(3)	ABANDON SHIP		(8)	FOLD/UNFOLD	
(4)	ALARMS, 1MC				
(5)	LIFEBOAT OPERATIONS				

Figure C-1. Sample format for a record of certification form (continued)

C-5. AMMUNITION SHIPBOARD COMPATIBILITY

Figure C-2 shows the types of ammunition used by Army aircraft and their compatibility with ships.

	BGM-71 (TOW) ARMY/NAVY	BGM-71(I-TOW) ARMY/NAVY	BGM-71(ER-TOW) ARMY/NAVY	BGM-71 (TOW2) ARMY	BGM-71 (TOW-2A)NAVY	BGM-71 (TOW2B) ARMY	AGM-114A (HELLFIRE) ARMY	AGM-114B (HELLFIRE) NAVY	AGM-114C (HELLFIRE) ARMY	AGM-114F (HELLFIRE) ARMY	AGM-114K (HELLFIRE II) NAVY	ATA STINGER	70MM MK66 MOD 2 ROCKET	70MM MK66 MOD 3 ROCKET	20MM HEI	30MM HE-DP	7.62MM MIX	50 CAL [illegible]
UH-1	N	N	N	N	N	N	N	N	N	N	N	N	N	N	N	N	Y	Y
UH-60	N	N	N	N	N	N	2	3	2	2	3	N	N	N	N	?	Y	Y
OH-58D(I)	N	N	N	N	N	N	2	3	2	2	3	Y	Y	Y	N	N	N	Y
AH-1F	2	2	2	2	2	2	N	N	N	N	N	N	Y	Y	Y	N	N	n
AH-64	N	N	N	N	N	N	2	3	2	2	3	Y	Y	Y	N	?	N	N

Y COMPATIBLE
N NOT ON AIRCRAFT
1 REQUIRES MAJOR SOFTWARE/HARDWARE CHANGES
2 NOT SHIPBOARD COMPATIBLE
3 COMPATIBLE, NEEDS FLIGHT CLEARANCE

C-6. STRIKEDOWN AND RECOVERY GUIDE

This paragraph will be completed in the next version of this manual. The matrix on the following page will be completed with either a "yes" or "no" in each of the boxes. The matrix will indicate if a particular ammunition is cleared for loading, strikedown, or download, and recovery on the ship. The Navy is compiling data for this matrix as this manual is being completed. Figure C-3 is a sample format for an Army ammunition and recovery guide.

WEAPON	HANGAR DECK		RECOVERY	
	LOAD	STIRKEDOWN/ DOWNLOAD	UNEXPENDED	HUNG
BGM-71 (TOW) ARMY/NAVY				
BGM-71 (I TOW)ARMY/NAVY				
BGM-71(ER-TOW) ARMY/NAVY				
BGM-71 ([illegible]				
BGM-71 (TOW2A) NAVY				
BGM-71 (TOW2B) ARMY				
AGM-114A (HELLFIRE) ARMY				

AGM-114B (HELLFIRE) NAVY				
AGM-114C (HELLFIRE) ARMY				
AGM-114F (HELLFIRE) ARMY				
AGM-114K (HELLFIRE) NAVY				
ATA STINGER				
MK 66 MOD 2 ROCKET				
MK 66 MOD 3 ROCKET				
M 267 MPSM WARHEAD				
M 274 PD WARHEAD				
20MM HEI				
30MM HE-DP				
7.62MM MIX				
.50 CAL MIX				

Figure C-3. Army ammunition and recovery guide

APPENDIX D

OPERATIONS FROM SINGLE- AND DUAL-SPOT SHIPS

This appendix provides information and procedures for flight deck operations on board several classes of ships not covered in the body of this manual. The procedures are intended for use on Naval vessels that OH-58D(I) and other detachments are commonly deployed on for missions such as those described below. NAEC-ENG-7576 is the official document for dimensions and landing capabilities of Navy ships. The diagrams included in this appendix are for planning only; they do not replace the shipboard resume.

D-1. OVERVIEW

a. In 1987, the US Army was directed to assume an overwater mission based on the Army's mission capability of night reconnaissance and security. Initially, special operations aircraft were deployed to the Arabian Gulf for Operation Earnest Will. This mission was to provide maritime reconnaissance and escort for oil tankers. Early in the mission, a unit of OH-58Ds was assembled, trained, and deployed to Operation Prime Chance, a mission they conducted very successfully.

b. In the years since Prime Chance, the Army has sustained an OH-58D(I) Kiowa Warrior overwater capability. Reasons for the Kiowa Warrior's popularity include its advanced night and stealthy operations capabilities. The OH-58D(I) is one of the few remaining aircraft that can land on specific-surface combatant ship frigates (FFG-7s). Since one-third of the ships in the Navy's Western Hemisphere Group are frigates, Army aviation can expect more opportunities to plan and participate in overwater operations such as sealane surveillance and small boat interdiction.

c. In addition to attack and reconnaissance operations, DOD guidance identifies the Army as the executive agent for battlefield medical evacuation, including the shore-to-ship mission. This indicates MEDEVAC aircraft will be called upon to conduct shipboard operations during future conflicts.

D-2. STANDARD SHIPS

a. The term, "standard ships," refers to the classes of ships that detachments are commonly deployed upon. Standard ships include--

- Oliver Hazard Perry-class guided missile frigates (FFGs 7 through 61).
- Spruance-class destroyers (DDs 963 through 992 and 993 through 996).
- Kidd-class guided missile destroyers (DDGs 993 through 996).

b. Figures D-1 through D-4 on pages D-3 through D-6 show typical deck dimensions for the flight decks of each of these ships.

D-3. NONSTANDARD SHIPS AND PLATFORMS

Occasionally, detachments are required to conduct operations from vessels or platforms that are not covered in this appendix. The procedures explained in this appendix should be used if they are suitable for that particular vessel or platform. If a detachment is to be deployed to a nonstandard vessel or platform for an extended period, flight deck procedures should be written and maintained on that platform and used as a unit SOP.

D-4. AIRCRAFT HANDLING

a. *Preventive Maintenance.* For deployed detachments, aircraft preventive maintenance will be performed in the ship's hangar after the flight or before preflight. When aircraft are in the hangar and before any maintenance is performed, the following steps normally will be taken. The unit SOP should specify safety procedures. Depending on the type of aircraft involved, this may include pulling certain circuit breakers, disconnecting the aircraft battery, or performing certain armament procedures. The unit needs to confirm these procedures at the presail conference.

b. *Preflight.* The preflight will be performed in the hangar before the mission briefing or before the aircraft are pushed out.

c. *Aircraft Push-Out.* Normally, the aircraft are pushed out onto the flight deck at least 30 minutes (45 minutes is optimal) before the scheduled takeoff time. Aircraft will not be pushed out if the ship is maneuvering, the weather precludes flying, or the sea state will not allow for safe ground handling.

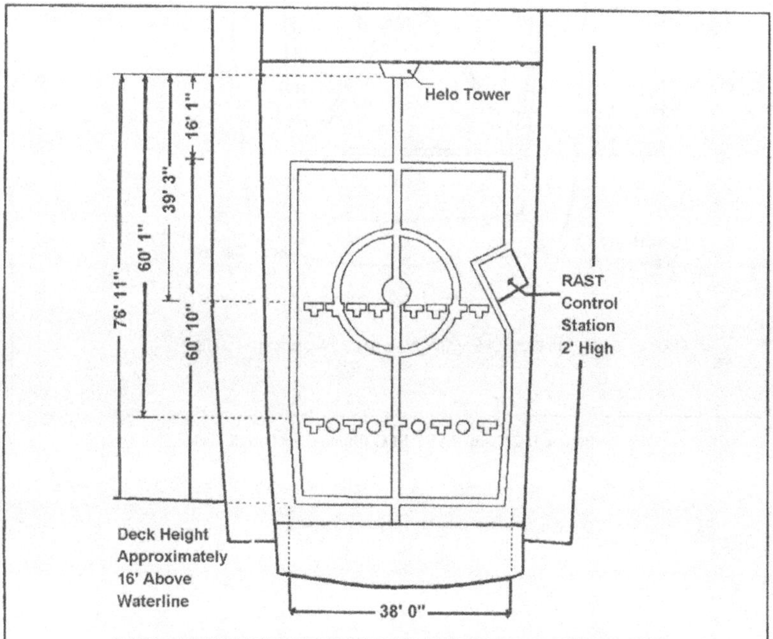

Figure D-1. RAST-modified FFG dimensions (large deck)

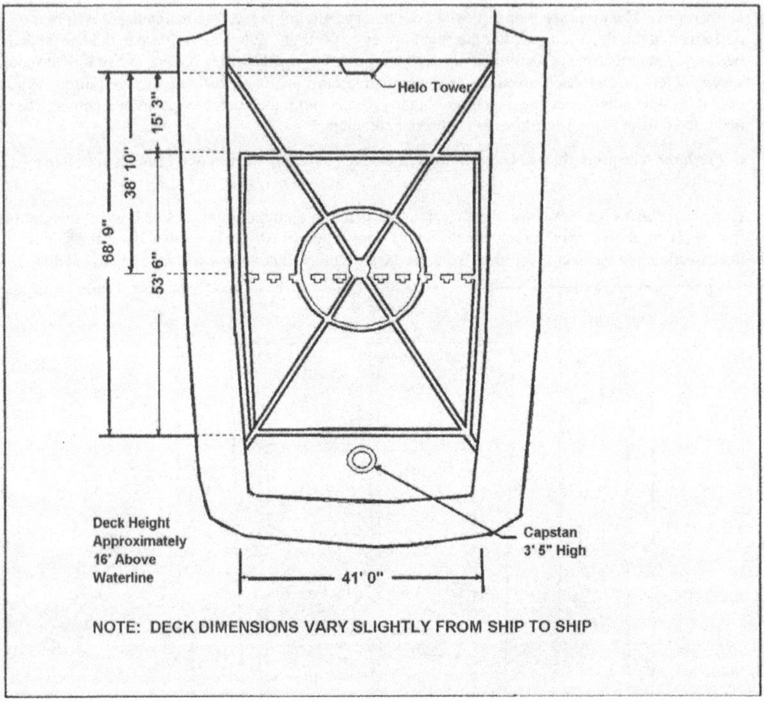

Figure D-2. Non-RAST FFG dimensions (small deck)

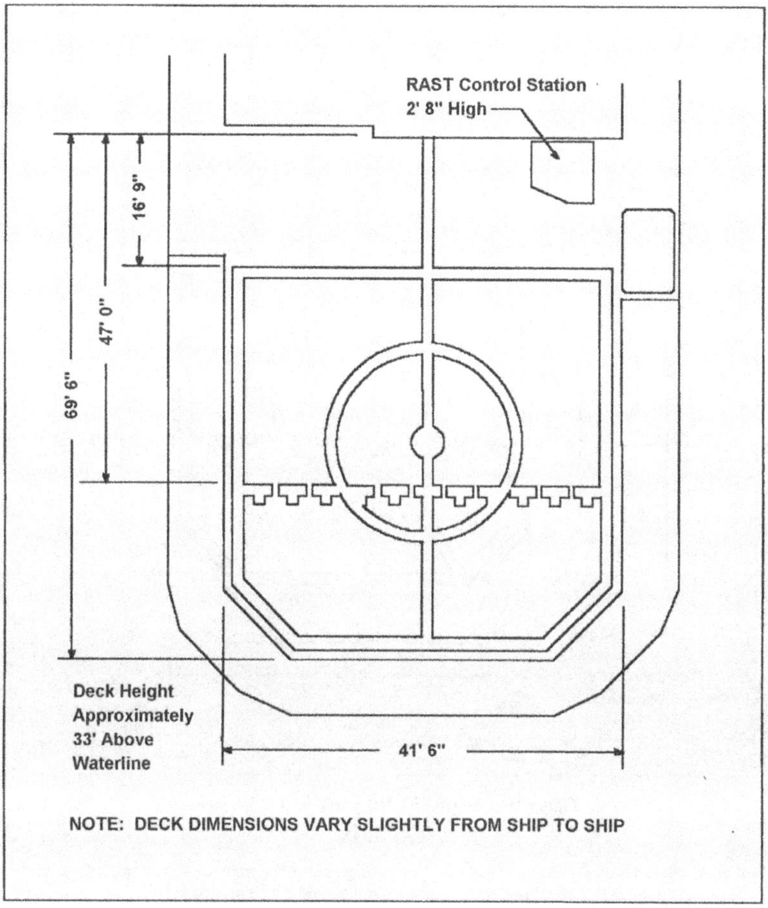

Figure D-3. RAST-modified DD/DDG deck dimensions

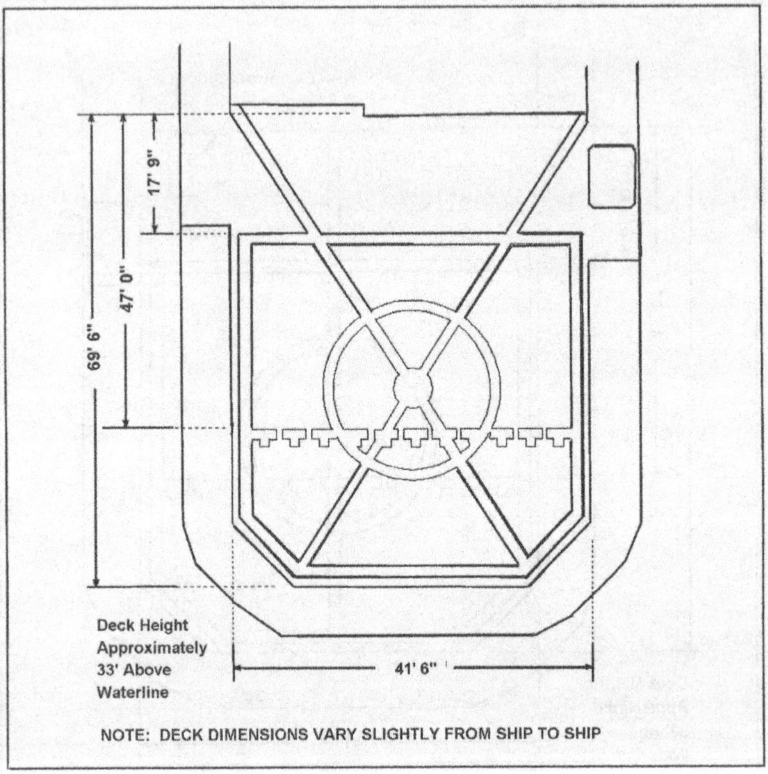

Figure D-4. Non-RAST DD/DDG deck dimensions

CAUTION

1. Before the aircraft are pushed out, contact the bridge to get clearance from the OD to move the aircraft. If pitch and roll is more than 2 and 4 respectively, the aircraft should not be pushed out.

2. Aircraft will be pushed out of or into the hangar with one man on each ground- handling wheel ready to release the hydraulic pressure in case of a runaway aircraft. A minimum of six personnel will be available to ground handle the aircraft.

d. ***Aircraft Spotting.*** When two OH-58D(I) aircraft are being moved from the hangar to the flight deck, the lead aircraft will be spotted and chained on the aft spot and the trail aircraft on the forward spot. Except for an emergency launch, all aircraft will be chained on opposite corners using at least two chains. One chain will be behind the .50-caliber machine gun (or other left-side weapon) and the other in front of the 2.75 FFAR rocket pod (or other right-side weapon), as shown in Figure D-5. If the pitch and roll of the ship is 2 and 4 respectively (or greater), four chains will be used. The following procedures will be adjusted to allow for the extra chains without compromising the safety of the deck crew. Personnel will not cross under the tail boom of a running aircraft or work under the tail boom except as described in paragraph h below. The PCs will stay with their aircraft when the rotor blades are being unfolded and visually ensure that flyaway items that were not removed during preflight are removed and taken away. These items will include the hub lock, blade rack, blade-folding wand, and any tools used during the unfolding process. The flyaway gear will be placed on the hangar floor and inventoried by the NCOIC before the engine is started. The rockets will be seated when the aircraft is spotted.

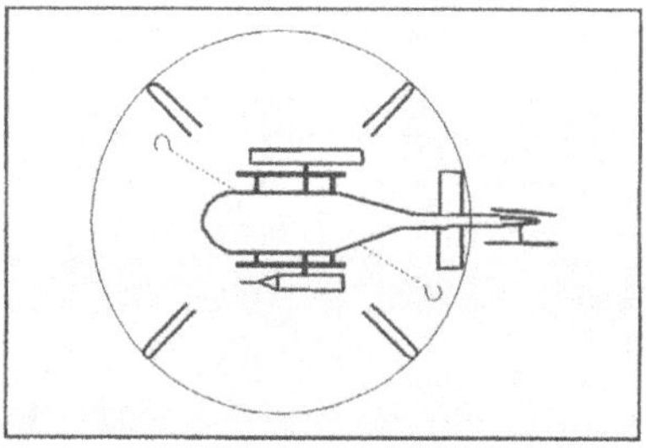

Figure D-5. Chain locations

e. *Engine Start and Run-Up.*

(1) *Required personnel.* Personnel on deck during engine start and run-up are limited to the following:

(a) *Designated LSE.* The LSE will maintain contact with the tower, control access to the flight deck, and monitor the trail aircraft.

(b) *Detachment NCOIC.* The detachment NCOIC is the safety observer and deck crew supervisor.

(c) *Detachment armament specialist.* The detachment armament specialist arms and dearms weapon systems.

(d) *Two detachment crew chiefs.* The detachment crew chiefs are the chain handlers.

(2) *Deck crew positions.* The flight deck crew will be positioned as described below. (See Figure D-6.)

(a) *LSE.* The LSE will be posted by the door to the hangar. He keeps track of the number of chains taken in and out of the hangar and the individuals who go in and out of the hangar. He will not allow anyone who is not involved with deck operations to leave the hangar and go onto the flight deck. Until the NCOIC is in a position to monitor the aircraft, the LSE also monitors the trail aircraft for light signals that indicate a need for assistance. If assistance is required, the LSE moves to the hangar side crew member, ascertains the nature of the problem, and notifies the appropriate maintenance person.

(b) *NCOIC.* The NCOIC is positioned outside the hangar-side crew member's door of the lead aircraft, clear of the weapon system on that side of the aircraft. He monitors the lead aircraft for light signals that indicate the need for assistance and monitors the activities of the deck crew.

(c) *Armament specialist.* The armament specialist is positioned beside the .50-caliber machine gun on the lead aircraft. The .50-caliber machine gun and the Aim 1/DLR laser will not be armed until the chain handler has passed in front of the machine gun on his way to the hangar with the chain from the stern side of the lead aircraft.

(d) *Chain handlers.* The chain handlers will be positioned by each aircraft on the side of the aircraft that is farthest from the hangar.

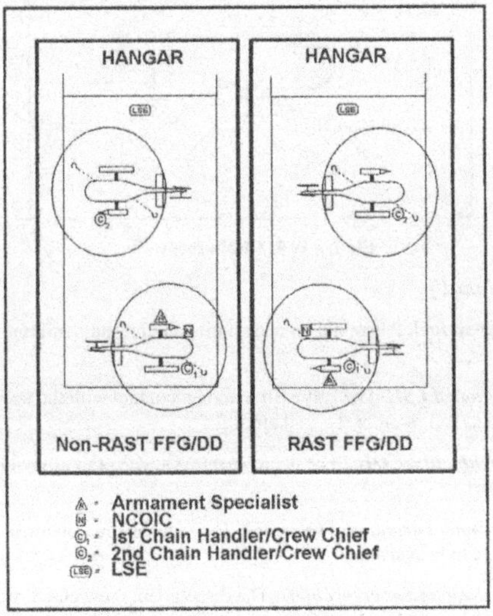

Figure D-6. Crew deck positions

(3) *Procedures.*

(a) The pilots will wait for an "amber deck" and use the aircraft checklist to start the engines. After the aircraft are started, the team should perform a communications check on the FM 1 secure radio. The pilot of the lead aircraft should contact the ship's control on UHF secure and request the ship's current position and course and speed for navigation alignment. He then ensures that the pilot of the trail aircraft has the information. If either of the aircraft experience maintenance problems, the pilot will signal the maintenance crews by flashing a flashlight or map light. Anytime the aircraft are running on deck at night, the aircraft position lights will be on and stay on until just before takeoff. Aircraft will stay chained with the rockets seated and the .50-caliber machine gun unarmed until the pilot receives a "green deck."

(b) When the pilots of both aircraft have completed navigation alignment and the systems are normal, the pilot of the lead aircraft will request a "green deck" from the tower. When the teams have been given a "green deck," the aircrews will signal the deck crews with a lateral wave of a flashlight or map light to remove the chains. (During day operations, a stroking of the arms with the hands is used.)

(c) The terms listed below are used with deck clearances:

• Amber Deck. Request to turn rotors or shut down aircraft on the deck; fire party and tower personnel must be present.

• Green Deck. Request to take off or land; the ship is not maneuvering and the bridge is aware of aircraft operations.

• Red Deck. The deck is not safe for operations.

f. ***Arming and Unchaining.***

(1) When a "green deck" is received, the chain handler removes the chain on his side of the aircraft and walks around the nose of the aircraft to the chain release on the other side. After the chain handler has passed in front of the .50-caliber machine gun, the armament specialist may arm the machine gun and the Aim 1/DLR laser (if installed) as assisted or directed by the crew of the lead aircraft. During the arming process, the chain handler may remove the remaining chain, show both chains to the NCOIC and pilot and--mindful of the rotor system--proceed around the nose of the trail aircraft and into the hangar. The chain handler will check in with the LSE upon entering the hangar (Figure D-7). When the lead aircraft is armed, the armament specialist will move to the .50-caliber machine gun and Aim 1/DLR laser of the trail aircraft--being mindful of the rotor systems. The .50-caliber machine gun and Aim 1/DLR laser of the trail aircraft will not be armed until the chain handler and NCOIC have passed clear of the front of the weapon system and are going toward the hangar.

(2) The NCOIC will visually ensure that the chains are removed, arming procedures are complete, and personnel are clear of the aircraft. He then taps the nearest pilot on the shoulder and gives him a thumbs up, indicating that all pretakeoff ground checks are complete. The NCOIC will remain at the lead aircraft until the pilot responds with a thumbs up. The NCOIC positions himself outside the crew member's door of the trail aircraft on the side farthest from the hangar and clear of the weapon system on that side. The NCOIC remains in a position to monitor both aircraft for light signals, which indicates a need for assistance. When the NCOIC is in position by the trail aircraft, arming and unchaining of the aircraft closest to the hangar may begin.

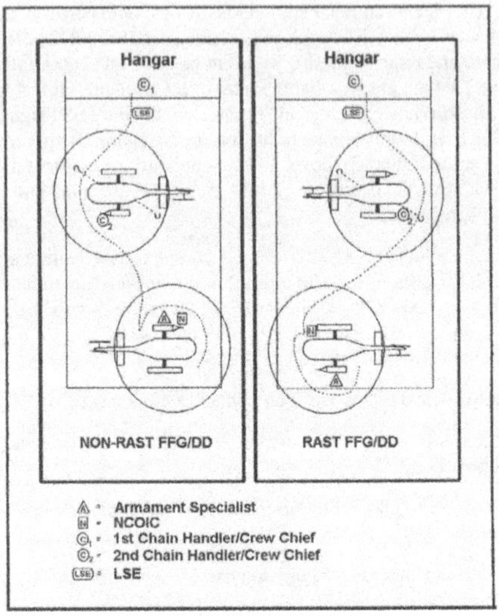

Figure D-7. Unchaining and arming the lead aircraft

(3) The second chain handler removes the chain from the stern side of the trail aircraft and, followed by the NCOIC, passes in front of the aircraft. The chain handler moves to the remaining chain release, and the NCOIC positions himself outside the crew member's door nearest the hangar and clear of the weapon system on that side of the aircraft (Figure D-8). When the chain handler and the NCOIC have passed clear of the .50-caliber machine gun, the armament specialist may arm the machine gun and the Aim 1/DLR laser. During the arming process, the chain handler may remove the remaining chain, show both chains to the NCOIC and pilot, and stow the chains in the hangar. Before entering the hangar, the chain handler will check in with the LSE.

(4) The armament specialist will arm the .50-caliber machine gun and Aim 1/DLR laser of the trail aircraft and proceed into the hangar, checking in with the LSE on his way in.

(5) The NCOIC will visually ensure that the chains are removed, arming procedures are complete, and personnel are clear of the aircraft. He then taps the nearest pilot of the trail aircraft on the shoulder and gives him a thumbs up, indicating that all pretakeoff ground checks are complete. The NCOIC will remain at the trail aircraft until the pilot responds with a thumbs up. The NCOIC then departs the flight deck and checks in with the LSE. The LSE follows the NCOIC into the hangar.

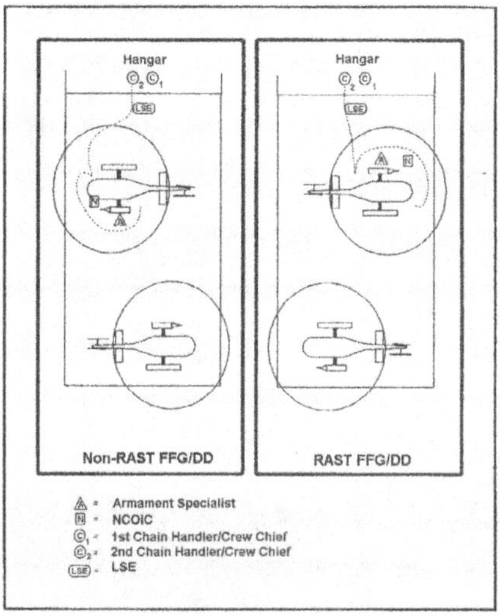

Figure D-8. Unchaining and arming trail aircraft

g. ***Dearming and Chaining Procedures.***

(1) This procedure will be reversed when recovering aircraft. It is simplified by dearming and chaining only one aircraft on deck at a time. Both aircraft will be dearmed and chained in such a way that the armament specialist dearms the .50-caliber machine gun and the Aim 1/DLR laser before anyone passes in front of the weapon system. The NCOIC's position will be such that he can monitor the dearming and chaining process on each aircraft. The LSE will allow deck crew personnel onto the flight deck only after he receives the appropriate light signal from the aircraft and the position lights are turned on.

(2) When the trail aircraft is secured, the LSE will follow the chain handler, armament specialist, and NCOIC into the hangar. Then the aircraft position lights are turned off. Personnel will remain in the hangar until the lead aircraft lands. The lead crew will request chains through the trail crew and the LSE or by turning on the position lights. If the trail aircraft is still running, the trail crew will relay the appropriate light signal to the LSE observing through the hangar door view glass and turn the trail aircraft position lights back on before deck personnel exit the hangar. The lead aircraft will be dearmed and chained the same way as the trail aircraft. No personnel will pass in front of the aircraft until the armament specialist has completed dearming the .50-caliber machine gun and Aim 1/DLR laser. While the armament specialist is dearming the machine gun and the laser, the chain handler may chain the aircraft on the side nearest the hangar. The chain handler will not pass in front of the aircraft until the armament specialist has completed the dearming procedure.

h. *Out-of-Position Aircraft.* During recovery, deviation from this procedure may become necessary if aircraft are out of position because of pitch and/or roll, the sea state, or adverse winds. If this occurs, deck personnel may pass under the tail boom if the following safeguards are maintained:

(1) The NCOIC positions himself at the horizontal stabilizer of the out-of-position aircraft. He will maintain contact between his shoulder or upper arm and the stabilizer.

(2) The NCOIC directs personnel under the tail boom between himself and the point where the tail boom is attached to the aircraft.

(3) If both aircraft are out of position, the chain handler and armament specialist will follow the NCOIC to the lead aircraft when they are clear of the tail boom of the trail aircraft. They will be cleared under the tail boom of the lead aircraft the same way. The procedure will be reversed upon returning to the hangar.

i. *Ground Safety Checks.* The personnel listed will complete the ground safety checks listed below.

(1) Flyaway gear is removed, stowed, and accounted for.

- Respective aircraft PC (during removal).
- Designated person inside the hangar (LSE).
- NCOIC (final check before the engine is started).

(2) Chains are removed, stowed, and accounted for.

- NCOIC.
- PC.
- LSE.

(3) Personnel are clear of the aircraft and flight deck.

- NCOIC.
- PC of the trail aircraft (relayed to lead).
- LSE.

j. *Hangaring Procedures for FFG and DD/DDG.*

(1) When aircraft have been refueled, dearmed, washed, flushed, a maintenance operational check conducted, and the blades folded, they are ready to be hangared.

NOTE: Before any aircraft are moved, a detachment member will contact the bridge and alert them that aircraft will be moved into the hangar. This limits the movement of the ship and allows for easier aircraft handling.

(2) The first aircraft into the hangar should be pushed in nose first closest to the centerline passageway. The aircraft will be pushed as far forward as possible and the tail brought toward the centerline passageway. The second aircraft goes in tail first on the outboard side of the hangar. When the aircraft are positioned in the hangar, a minimum of four chains will be used to secure them to the deck. To limit aircraft movement during rough seas, more chains may be necessary from the side of the aircraft to the bulkhead.

D-5. WEATHER REQUIREMENTS

This paragraph is included as an example for developing unit SOPs on weather during shipboard operations.

a. ***Overland Flight Operations.*** All flights conducted over land will comply with current published weather minimums.

b. ***Overwater Flight Operations.*** Because of the absence of weather reporting facilities at sea and the unique complications of NVG flight over water, weather minimums become more subjective than over land. When faced with making a weather decision before each flight or during flight, the senior officer will consider the opinion of each crew member in the final decision.

(1) ***Overwater flights during the day.***

- The ceiling and visibility will be according to current published weather minimums.
- A visible horizon is not required if VMC is maintained.

(2) ***Unaided overwater flights at night.*** When necessary or required, weather minimums and the required equipment will be according to current regulations.

(3) ***Overwater flights using NVG.***

(a) If moon illumination is equal to or greater than 23 percent and is 30 degrees or more above the horizon--

- The ceiling and visibility is according to current published weather minimums.
- A visible horizon is required in a minimum of three quadrants (270 degrees continuous NVG-visible horizon).

(b) If moon illumination is less than 23 percent or less than 30 degrees above the horizon--

- The ceiling and visibility is according to current published weather minimums.
- A visible horizon is required in all quadrants (360 degrees continuous NVG-visible horizon).
- The minimum sea state will be 1.

D-6. MAXIMUM WINDS

This paragraph contains diagrams showing the maximum allowable relative winds for taking off and landing on single- and multiple-spot flight decks.

a. ***Preferred Winds.*** The preferred maximum winds for taking off and landing on all single- and multiple-spot decks is 10 knots or less of relative wind from any direction.

b. ***Wind Diagrams.***

(1) The wind diagram shown in Figure D-9 on page 17 is used when two OH-58D(I)s are positioned on a single-spot flight deck at the same time. This includes situations when the blades of one aircraft are folded and the second aircraft is taking off or landing or when a single aircraft will be flown off or landed to the opposite corner (forward or aft) deck positions.

(2) The wind diagram in Figure D-10 on page 18 is used when a single OH-58D(I) is flown off or landed to the center position of single-spot flight decks.

(3) The wind diagram in Figure D-11 on page 19 is used when aircraft are taking off or landing to multispot flight decks.

c. ***Deviations.*** The wind diagrams discussed in this paragraph are intended to be the maximum allowable for normal operations; however, the PC will make the final determination.

D-7. TAKEOFF PROCEDURES

a. So that enough time is allowed for all systems to become operational, flight quarters should be set 30 minutes before the scheduled takeoff time. This allows enough time for the fire crew and tower personnel to man their stations. This time must be coordinated with the ship's crew (LAMPS) and the TAO to help reduce the amount of time that Navy personnel are on flight quarters. If the ship is equipped with a RAST, LAMPS personnel will be responsible for moving it before the aircraft are pushed out. An "amber deck" should be requested 15 minutes before takeoff. To conserve battery power, the LSE, who has communications with the tower, will request the "amber deck."

b. When all systems are operating normally on both aircraft, there is a "green deck," the weapons are armed, the chains are removed, and all deck crew members are clear of the flight deck, the aircraft on the stern of the deck (lead) will take off to port or starboard, depending on the type of ship. (See Figure D-12 on page 20 and Figure D-13 on page 21.) A hover power check should be performed before leaving the deck to ensure that enough power is available. When the stern aircraft is clear of the deck, the pilot will call "OPS NORMAL" and give the "STATE" to the tower. (The state is endurance with current fuel in hours and minutes.) When the pilot of the forward aircraft hears that the stern aircraft is "OPS NORMAL," he will take off. The pilot of the lead aircraft also will call for the formation to rejoin and give a heading on the internal FM secure radio. The pilot of the trail aircraft will call the lead aircraft and advise the pilot that the flight is formed when the aircraft are formed up. The pilot of the lead aircraft will not take any vectors until the pilot of the trail aircraft makes this call.

NOTE: The maximum winds for normal operations are given in paragraph D-6; however, the PC makes the final determination.

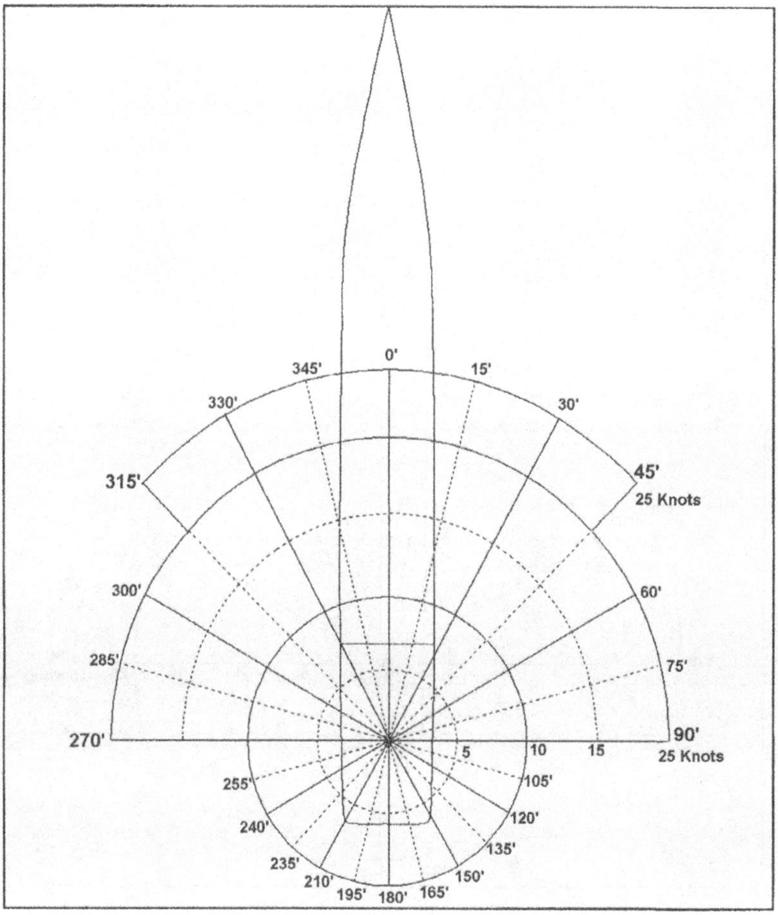

Figure D-9. Single-spot deck with two OH-58D(I)s positioned on opposite corners

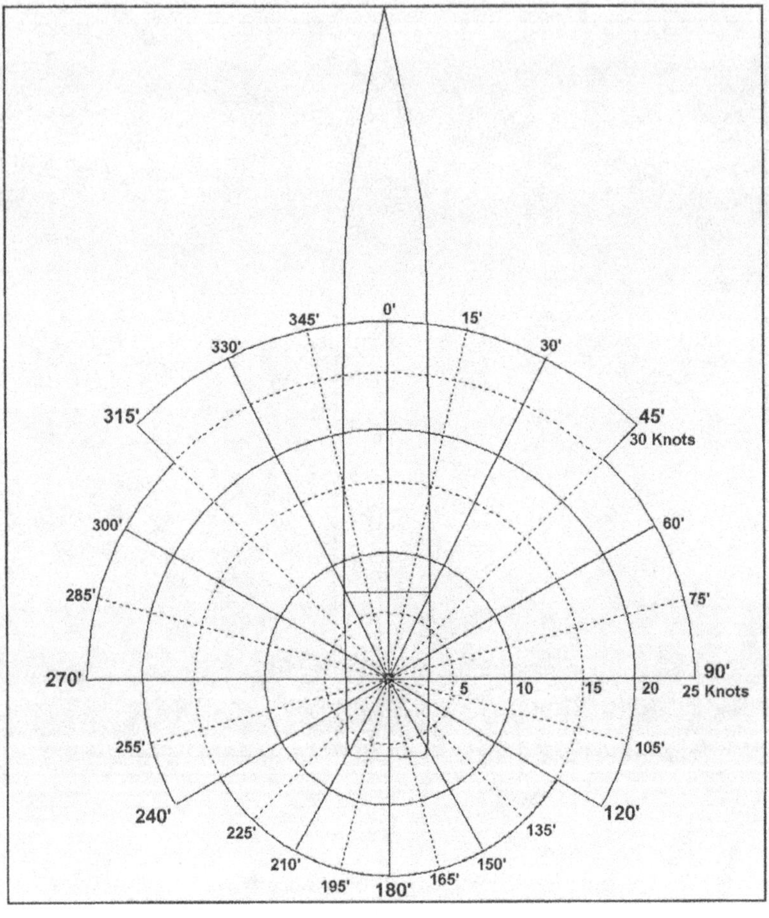

Figure D-10. Single-spot deck with one OH-58D(I) positioned on center spot

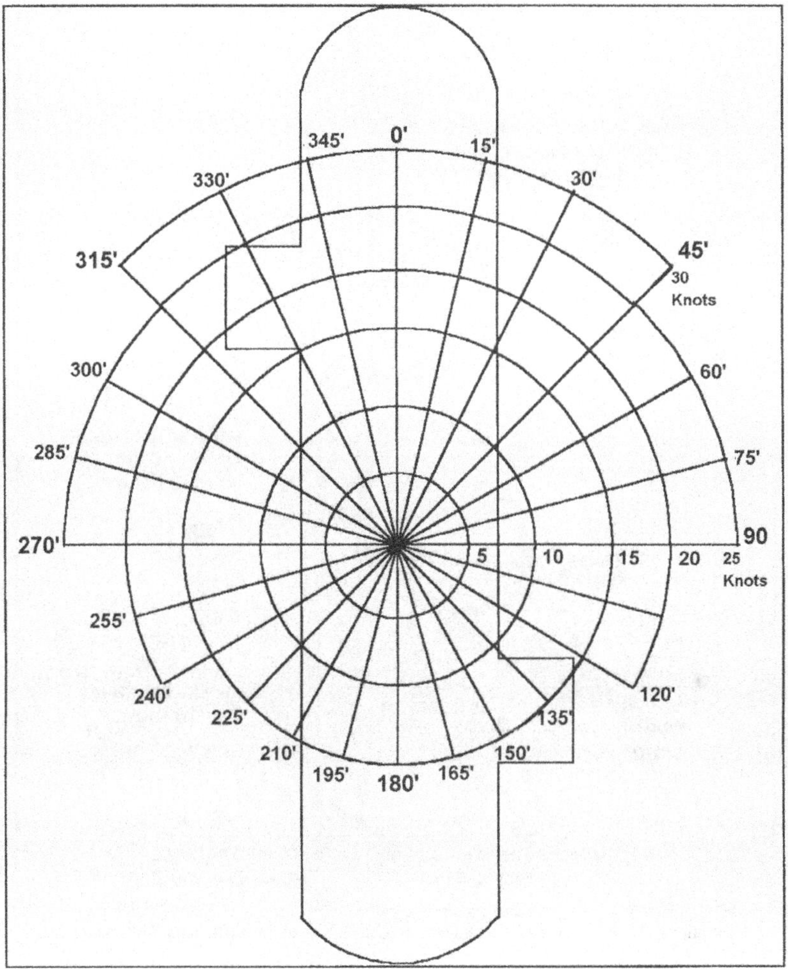

Figure D-11. Multispot decks (OH-58D(I) launch and recovery wind limits)

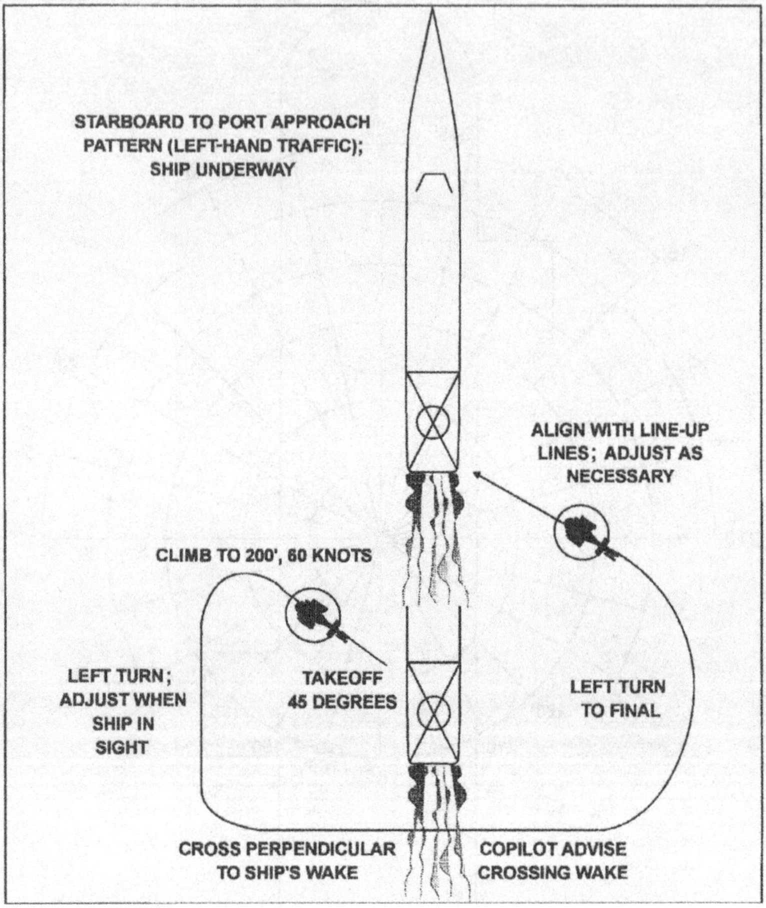

Figure D-12. Starboard-to-port approach pattern, left-hand traffic (ship underway)

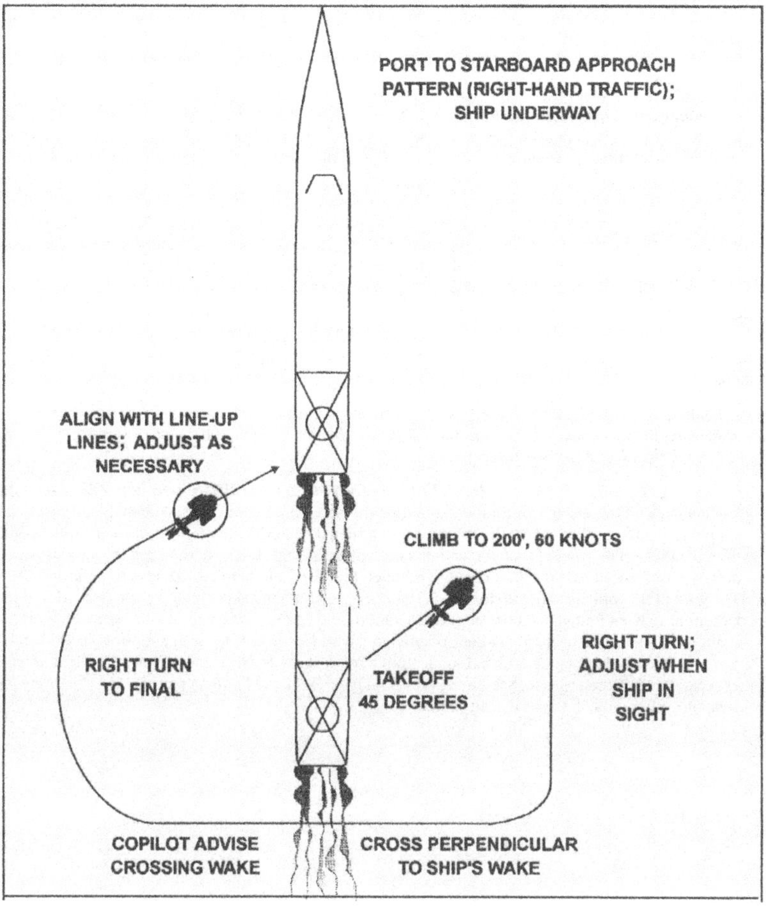

Figure D-13. Port-to-starboard approach pattern, right-hand traffic (ship underway)

D-8. RECOVERY

When the team completes a flight period, the ship's CIC should be informed as soon as possible so that personnel can be assembled for flight quarters. Normally, this requires 10 to 15 minutes. Once the flight is given a "green deck" for landing, the pilot of the aircraft will perform a before-landing check and ensure that the weapons are safed and ASE is turned off. When both aircraft are on deck and have been shut down and flushed and the blades folded, they can be washed. (Fresh water should be used daily; soap should be used every third flight.)

a. ***Landings on Large-Deck, RAST-Equipped FFGs (FFG 8 and 36 through 61).*** When the aircraft have been given a "green deck," the trail aircraft should break off from the formation and land port to starboard to the forward portion of the flight deck with the left skid parallel to the forward foul line. Once on the deck, the aircrew will signal the deck crew for dearming and chaining. When the weapons and chains are secured, the pilot of the trail aircraft will advise the pilot of the lead aircraft when the deck crew is clear and the lead is clear to land. The pilot of the lead aircraft should advise the tower when he is on short final and land starboard to port on the aft portion of the flight deck with the left skid parallel to the aft foul line (Figure D-14). The weapon systems may be dearmed as the aircraft are chained. The RAST should be positioned on the center of the deck.

NOTE:

1. With low or no illumination or when the ship's hangar shadows the deck, the IR searchlight should be used for landing.

2. When landing to a large-deck FFG (RAST-modified), one aircraft should land at a time when the relative winds are greater than 25 knots if either aircraft will have more than a 20-knot tailwind or if pitch or roll are equal to or more than 2 or 4 degrees respectively.

b. ***Landings on Small-Deck Non-RAST FFGs and Non-RAST DD/DDGs.*** Because of the size of the landing deck on these ships, one aircraft at a time should land. The trail aircraft should land first to help simplify operations for the next flight. The landing should be to the center of the deck or forward on the deck parallel to the forward foul line with the nose of the aircraft facing port. When the aircraft is on the deck, the aircrew signals the deck crew with a lateral wave of a light to chain and dearm the aircraft. The pilot of the trail aircraft will then request an "amber deck" and shut down. When the blades of the aircraft have been folded, the pilot of the lead aircraft will be cleared to execute his approach (port to starboard) to the stern of the flight deck parallel to the aft foul line. If the lead aircraft is fuel-critical, both aircraft may land to the deck before the trail aircraft shuts down. There will be only eight to ten feet of separation between rotors. (See Figures D-15 on page D-24 and D-16 on page D-25.) Dearming the weapon systems may be done during shutdown.

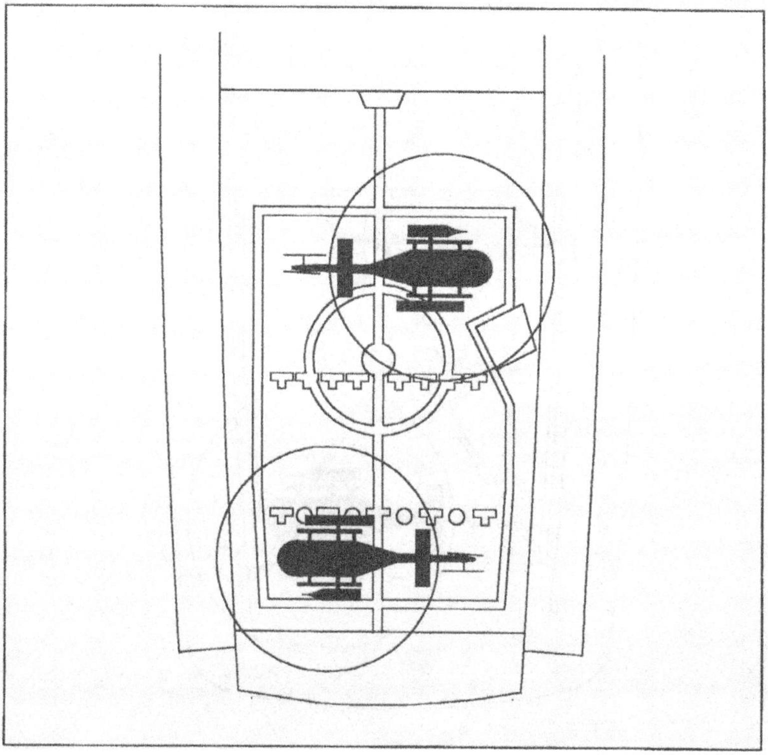

Figure D-14. RAST FFG takeoff and landing positions

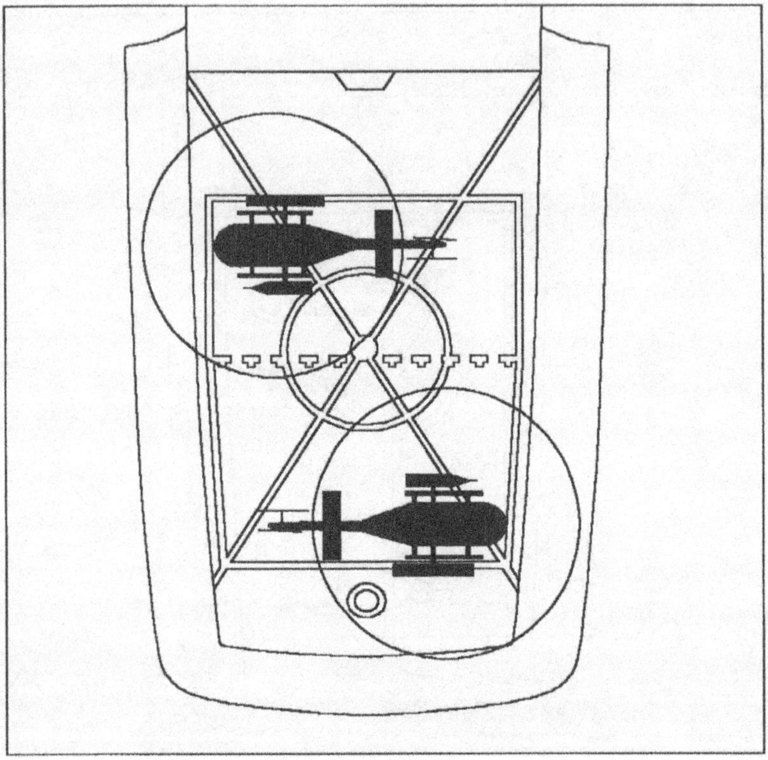

Figure D-15. Non-RAST FFG takeoff and landing positions

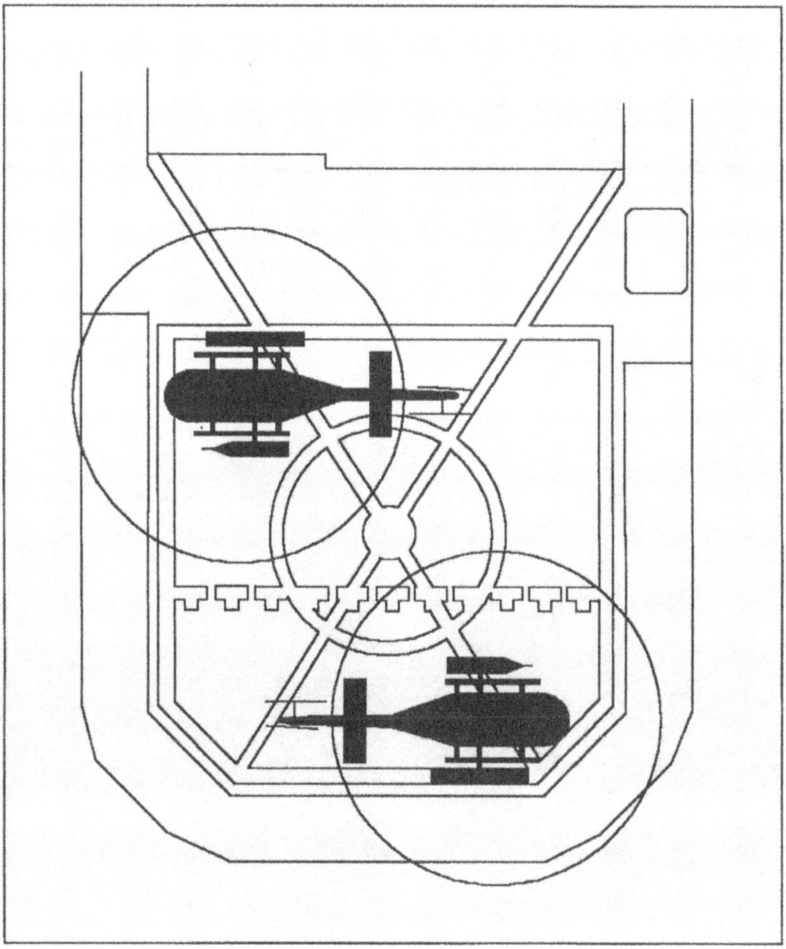

Figure D-16. Non-RAST DD/DDG takeoff and landing positions

NOTE: With low or no illumination or when the ship's hangar shadows the deck, the IR search light should be used for landing.

c. ***Landings on RAST-Equipped DD/DDGs.*** The flight deck of the RAST-equipped DD/DDG is approximately 6 inches narrower, 7 inches longer, and 17 inches closer to the hangar face or nearest obstruction than the non-RAST-equipped DD/DDG. Also, the POL island on the starboard side is moved aft about 3 feet. For these reasons, landing procedures to this deck should be similar to operations on other DD/DDG and small-deck FFG. However, the aircraft must be landed with the trail aircraft forward-most on the deck, the nose facing starboard, and the left-side weapon system about 17 inches aft of and parallel to the forward foul line. When the aircraft is chained down and shut down, its weapons secured, and the right-side blades folded, the lead aircraft can land to the aft position, nose to port. The aircraft will be positioned so that the left seat is aligned with the VERTREP marking (Figure D-17).

NOTE: With low or no illumination or when the ship's hangar shadows the deck, the IR searchlight should be used for landings.

d. ***Landing Fuel.*** The pilot will declare "minimum fuel" when 40 minutes of fuel remain as computed from the actual fuel flow. He will declare "emergency" when 20 minutes of fuel remain or the low-fuel caution light illuminates.

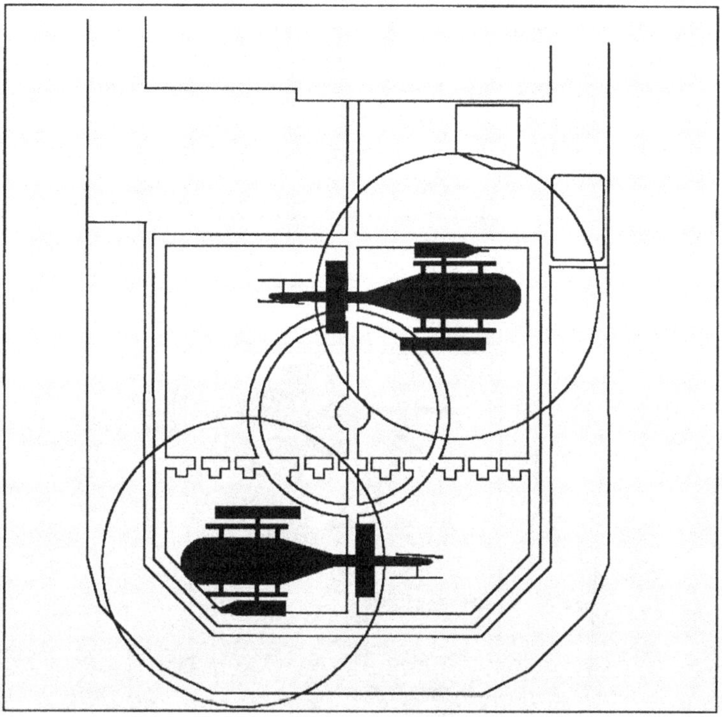

Figure D-17. RAST DD/DDG takeoff and landing positions

Figure D-17. RAST DD/DDG takeoff and landing positions

D-9. SHIPBOARD HOT REFUELING AND REARMING

a. ***Refueling and Rearming Procedures (Large-Deck FFG).***

(1) ***Two-aircraft operation (hot refueling).*** After both aircraft have landed in their normal positions, as shown in Figure D-18, refueling and rearming procedures will be conducted as described below.

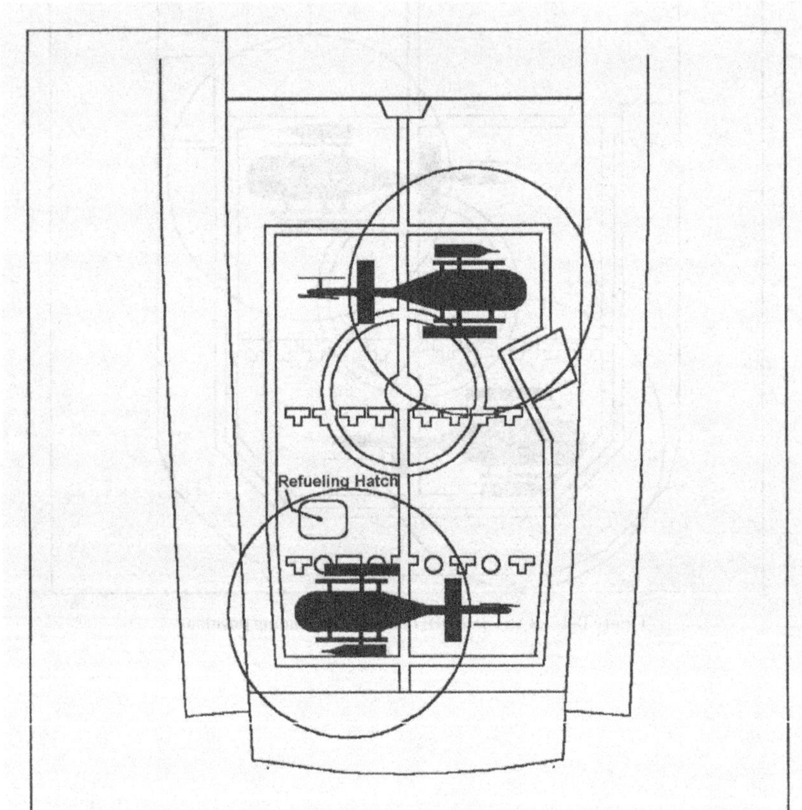

Figure D-18. RAST FFG hot rearming and refueling positions

(a) Fourteen rockets and 1,000 rounds of 50-caliber ammunition may be pre-positioned on the center spot for multiple loads.

(b) Aircraft armament systems will be safed before refueling. (Rocket tube igniter arms will be positioned in the up position and the .50-caliber switches in the safe position.)

(c) Stern aircraft should be refueled first because of the proximity of the refueling hatch on the deck.

(d) While one aircraft is being refueled, the other aircraft may be rearmed. The .50-caliber machine gun may be loaded in the can and chute, but it will not be fully armed until the aircraft are refueled and ready to take off. Rockets will not be loaded while the aircraft is being refueled, and the ignition arms will not be placed in the down position until the aircraft is ready to depart.

(e) After refueling is complete, the takeoff will be as described in paragraph D-7.

(2) ***Single-aircraft operation.*** The procedures for single-aircraft operation will be the same as for dual aircraft operations. However, the crew may land to the center of the deck, allowing room for access to the refueling hatch.

CAUTION

The ALQ-144 must be turned off before refueling begins.

(3) ***Cold refueling.*** The procedures for cold refueling are the same as previously discussed, but the aircraft are shut down and the blades must be at a complete stop. The blades may be folded during refueling and rearming procedures.

b. ***Rearming and Refueling Procedures (Small-Deck FFGs or DD/DDGs).***

(1) All hot rearming and refueling procedures should be performed single-ship. The starboard-to-port approach should be used, and the nose of the aircraft pointed to the port side of the hangar.

(a) ***FFGs.*** On FFGs, the POL point is located on the port side of the aft end of the deck (Figure D-19).

(b) ***DD/DDGs.*** On DD/DDGs, the POL point is located on the forward starboard side of the deck (Figures D-20 and D-21 on pages D-31 and D-32).

(2) The .50-caliber ammunition and rockets may be placed on the flight deck just outside the hangar door. The .50-caliber ammunition may be loaded into the can during the refueling operation, but it will not be armed until the refueling operation is completed. The rockets should be loaded when the refueling operation is completed. The igniter arms will not be placed in the down position until the aircraft is ready for takeoff. When rearming and refueling procedures are complete, the aircraft will depart and remain in the port or starboard deltas until both aircraft are ready. (The lead aircraft should land and be refueled first.)

WARNING

The ALQ-144 must be turned off before refueling begins.

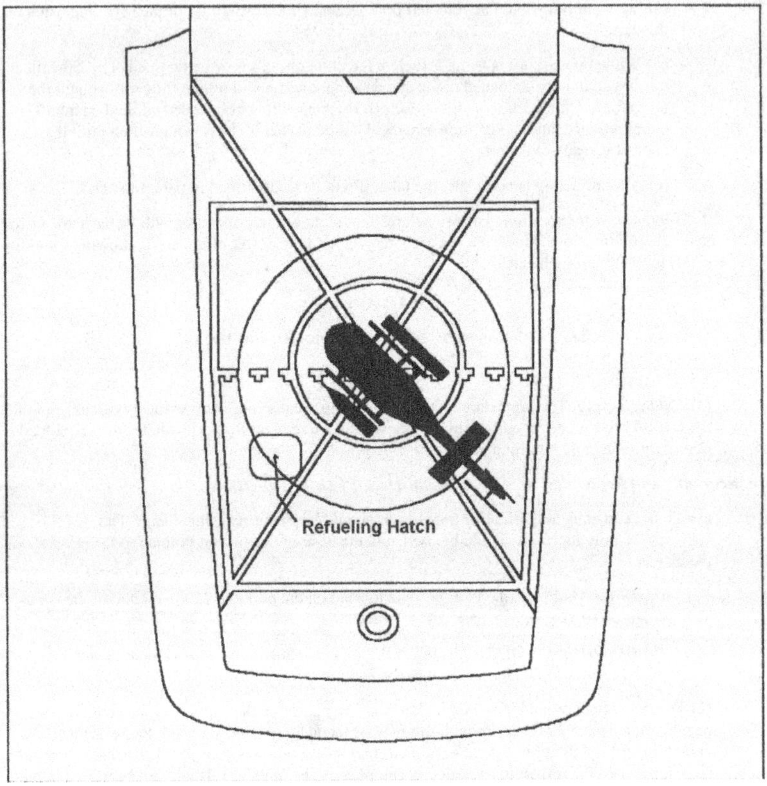

Figure D-19. Non-RAST FFG hot rearming and refueling position

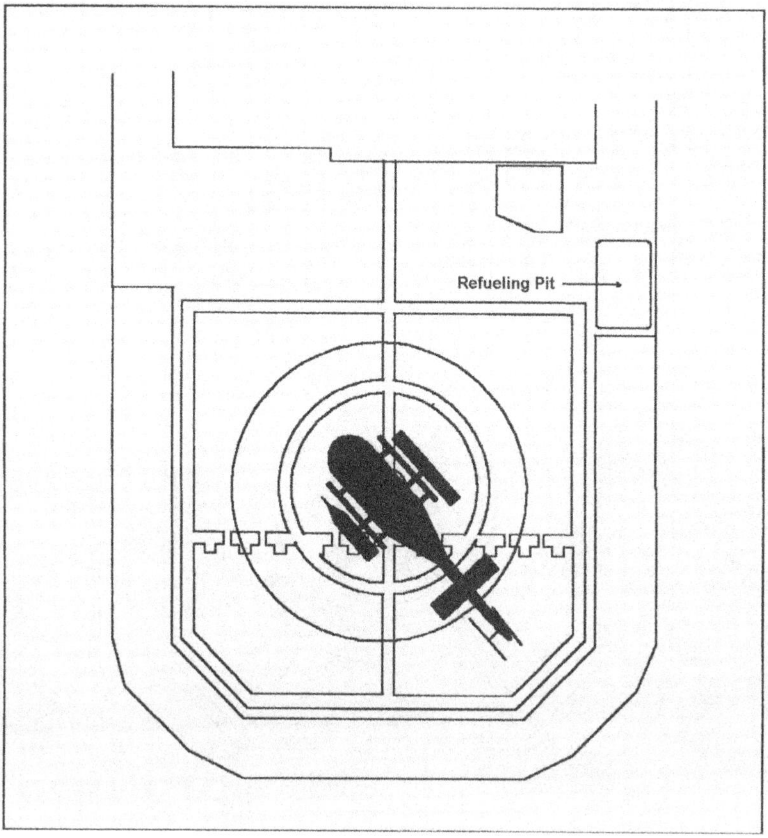

Figure D-20. RAST DD/DDG hot rearming and refueling position

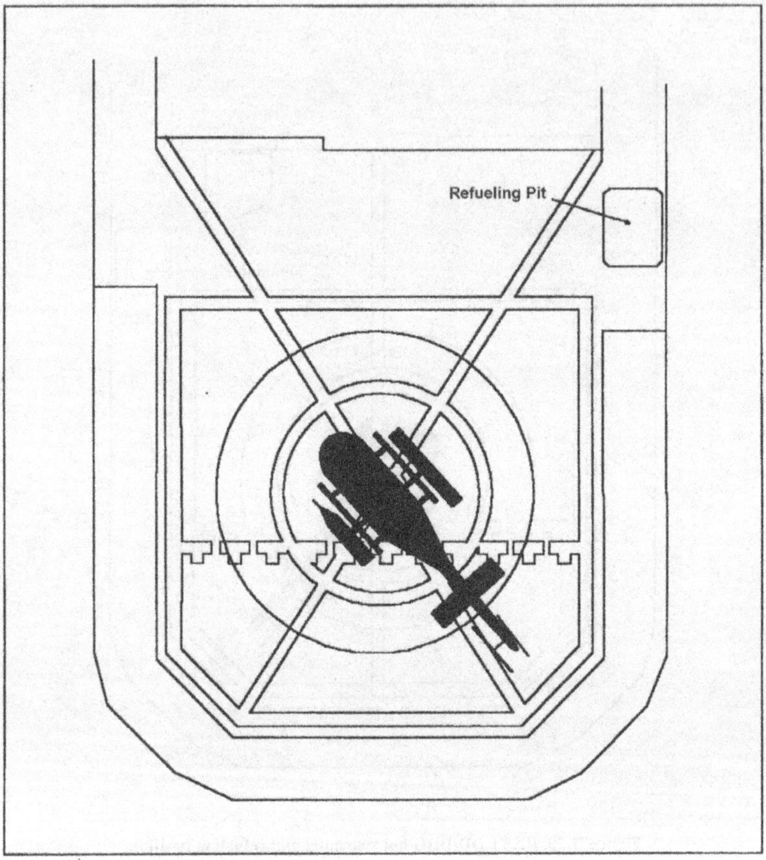

Figure D-21. Non-RAST DD/DDG hot rearming and refueling position

D-10. DECK OPERATIONS CHECKLIST

Figure D-22 is an example of a deck operations checklist.

1. AIRCRAFT MOVEMENT

a. RAST positioned.

b. Safety equipment (cranials, PFD, and lights) in place.

c. Clearance to move the aircraft granted.

d. All six personnel in position and briefed.

(1) One on each ground-handling wheel.

(2) One on each side of the horizontal stabilizer.

(3) One on the tail stinger.

(4) One ground guide giving the movement commands.

e. Chains off and reposition aircraft.

f. Chains on.

2. AIRCRAFT PREPARATION

a. Blades unfolded; blade rack stowed in the hangar.

b. Hub locks removed and stowed in the hangar.

c. Blade wands stowed in the hangar.

d. Ground-handling wheels removed and stowed in the hangar.

3. TI INSPECTION

a. Check main rotor blades and ensure that expandable bolts are installed and safed.

b. Hub locks, blade racks, ground-handling wheels, and blade wands removed and stowed.

c. Walk-around inspection completed.

4. AIRCRAFT LAUNCH

a. Request for amber deck acknowledged.

b. Clear deck of all nonessential personnel.

c. Aircraft started.

d. Rockets seated (as appropriate).

e. Request for green deck acknowledged.

f. Chains removed; gun armed and acknowledged by the pilot.

g. Deck cleared of ALL personnel.

h. Aircraft depart.

5. AIRCRAFT RECOVERY

a. Deck cleared of ALL personnel.

b. First aircraft lands (W #2).

c. Chains on when the pilot requests.

d. Weapon systems cleared.

e. Deck cleared of ALL personnel.

f. Second aircraft lands (W #1).

g. Chains on when the pilot requests.

h. Weapon systems cleared.

i. Aircraft shut down.

j. Engine flush as required.

Figure D-22. Example of a deck operations checklist

D-11. COMBAT SEARCH AND RESCUE

This paragraph provides general guidelines for combat SAR procedures.

a. *Search and Rescue Assets.*

• ***Primary.*** Naval or task force UH-60 helicopters are the primary SAR assets.

• ***Alternate.*** Naval surface vessels are the alternate SAR assets if a helicopter is not available.

• ***Backup.*** The OH-58D(I) is the backup SAR asset if the primary and alternate assets are not available for the SAR extraction.

b. *Situations.* On station, SAR operations fall into two basic categories. These categories are discussed in the paragraphs below.

(1) *In contact.* If an aircrew has been shot down by hostile fire, the PC of the remaining aircraft must evaluate the situation quickly. Ideally, the primary SAR asset will perform the SAR operation with the other OH-58D(I) providing cover. If the primary SAR is not available, the alternate SAR will be used. The OH-58D(I) will provide cover. When the primary and alternate SAR assets are not available or the situation calls for immediate extraction, the OH-58D(I) crew may opt to perform the SAR extraction.

(2) *No contact.* This situation allows for more options. The remaining OH-58D(I) crew should use the primary or alternate SAR as described above. However, the backup SAR should be used only in an emergency situation such as the threat of hypothermia.

D-12. INADVERTENT INSTRUMENT METEOROLOGICAL CONDITIONS

a. *Initial Actions.* When inadvertent IMC is encountered during formation flight, the most important consideration is aircraft control. Pilots must clearly understand the briefed IMC procedures to be executed. When pilots are deciding whether to perform IMC break-up, they should consider--

- The visibility.
- The ability of the trail aircraft to maintain formation integrity.
- The reliability of the navigational equipment in each of the aircraft.
- The type of navigation equipment in each of the aircraft.
- The expected recovery procedure.

b. ***IMC Break-Up Procedures.*** If inadvertent IMC is encountered during a standard staggered-right, two-aircraft team formation, the lead aircraft should climb straight ahead to base altitude. The second aircraft will turn 30 degrees away from the formation and climb to the base altitude plus 200 feet (Figure D-23). The base altitude must be established during the mission briefing based on known obstacles and the enemy situation. If more than two aircraft are in the formation, an alternate IMC break-up procedure must be briefed in detail.

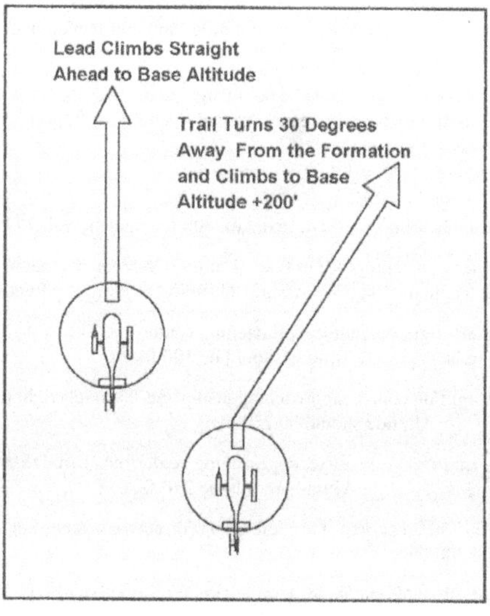

Figure D-23. Inadvertent IMC breakup

c. ***Recovery Procedures.*** Recovery procedures depend on the availability of radar and navigation equipment. When operating from a ship, the team may perform a TACAN approach or an emergency low visibility approach. (Paragraph D-13 discusses ELVA procedures and paragraph D-14 discusses TACAN procedures.) When operating under radar control of a LAMPS aircraft, the LAMPS may vector the team in the direction of known VMC. Another option is to perform an emergency IMC descent to VMC.

D-13. EMERGENCY LOW VISIBILITY APPROACH

a. The ELVA procedure is a nonprecision radar approach intended primarily for emergency recovery of aircraft experiencing inadvertent IMC. The approach is performed using the ship's fire control radar and is sometimes referred to as a carrier control approach.

b. Figure D-24 shows the radio calls that will be made by the ATACO during an ELVA approach.

1. This will be a radar-assisted approach. Have you radar contact on the _______radial, _______miles from the ship. Altimeter setting is _______. Weather is: ceiling _______, visibility _______. Final approach heading will be _______. Winds are _______degrees port/starboard at _______. Maximum pitch _______, roll _______. Read back altimeter setting.

2. Descend/climb to and/or maintain 400 feet. Assigned heading is _______.

3. Lost communications procedures are as follows: If no transmissions are received in one minute in the pattern or 15 seconds on final, climb to and maintain 400 feet. Attempt contact on secondary _______. If unable to make contact on secondary, squawk Mode 3, 7700 for 1 minute, then 7600. Alternate approach will be TACAN CH _______commencing at 3 miles and 400 feet on the _______radial. Acknowledge.

4. Missed approach procedures are: If ship or wake not in sight at missed approach point, turn left 30 degrees immediately, climb to 400 feet and increase airspeed to 80 knots. Report level and stand by for further instructions.

5. Perform landing check.

6. Turn right or left to the final bearing _______, maintain 400 feet and slow to 70 knots.

7. Do not acknowledge further transmissions. On final, 4 miles. Commence gradual rate of descent to arrive at 1/2 mile at 50 feet. Maintain 70 knots. Assigned heading is _______. Report "See Me."

8. (Call Sign) 3 1/2 miles. Left/ right/on course, approaching centerline. Turn left/right (corrective heading) or assigned heading is _______. Altitude should be 300 feet.

9. (Call Sign) 3 miles. Left/right/on course, approaching centerline. Turn left/right (corrective heading) or assigned heading is _______. Altitude should be 250 feet.

10. (Call Sign) 2 1/2 miles. Left/right/on course, approaching centerline. Turn left/right (corrective heading) or assigned heading is _______. Altitude should be 250 feet.

11. (Call Sign) 2 miles. Left/right/on course. Turn left/right (corrective heading) or assigned heading is _______. Altitude should be 200 feet.

12. (Call Sign) 1 1/2 miles. Left/right/on course. Turn left/right (corrective heading) or assigned altitude is _______. Altitude should be 150 feet.

13. (Call Sign) 1 mile. Left/right/on course. Turn left/right (corrective heading) or assigned altitude is _______. Altitude should be 100 feet. Slow to 40 knots.

14. (Call Sign) 1/2 mile; assigned heading is _______. Maintain 50 feet and 40 knots.

15. (Call Sign) 800/600/400/200 yards. Centerline is left/right straight ahead.

16. (Call Sign) at missed approach point. If ship or wake not in sight, execute missed approach.

Figure D-24. ELVA approach procedures

D-14. TACAN APPROACH

The TACAN procedure is a nonprecision approach intended to recover aircraft in IMC. This approach should be the alternate approach for aircraft experiencing inadvertent IMC. The approach will be used as a recovery means if the ELVA approach is unavailable.

a. Aircrews experiencing inadvertent IMC will declare an emergency, request the "180 relative approach," and request no holding at primary marshal. The 180 relative approach reduces the pilot workload by making the BRC the final approach course (BRC 325 = TACAN final approach course 315). This also increases the possibility of the aircraft establishing visual contact at the missed approach point due to full the flight deck lighting facing aft on most air-capable ships. Recommended airspeed on final for a TACAN approach is 60 knots.

b. In case a stuck card occurs on the OH-58D(I) TACAN BDHI, an alternate means of approach may be used. The pilots will place the TACAN BDHI bearing pointer on the fixed index and begin the standard TACAN approach at the 3 DME mark as indicated on the standard TACAN approach profile view. The crew must advise the ship of the inbound heading (from VSD) and request that the ship turn to that heading for the BRC. If the ship cannot turn, the pilots must realize the ship may appear at an oblique angle on break-out which increases the possibility of spatial disorientation during the transfer to visual references. The pilot not on the controls should monitor the attitude indicator closely for unlevel indications and be prepared to assume the controls.

D-15. LOST COMMUNICATIONS OR RADAR CONTACT

During overwater tactical operations, the team normally is under positive radar control and in radio contact with the controlling agency. If the flight is to be conducted under radio and/or radar contact and contact is lost, the flight should close in on the ship as soon as they realize the condition. At ranges beyond 12 nautical miles, difficulties with radio or radar contact may occur. Closing in on the ship should remedy the situation. The direction in which to fly to close in on the ship may be determined by--

- The last known direction to the ship.
- TACAN BDHI indications.
- Tracking toward AN/APR-39 signals from the ship's surface search or fire control radar.

a. ***Lost Communications.***

(1) ***Within the flight.*** If communications are lost between aircraft that are operating as a flight--

(a) Use light signals to convey the problem.

(b) Return to the ship with one aircraft relaying the problem. (The aircraft with communications problems will land first, and maintenance will begin troubleshooting the problem.)

(2) ***Between the flight and the ship.*** If communications are lost between the flight and the ship, the flight and the ship will follow the procedures given below.

(a) The pilot should squawk Mode 3/A code 7700 for 1 minute, then change to code 7600.

As the threat situation allows, he will repeat this procedure every 15 minutes for the duration of the flight.

(b) The aircraft and the ship should continue to make radio transmissions in the blind on the primary and alternate briefed frequencies.

(c) Upon arrival back at the ship, the flight should hold in port deltas or at a hover, if the power margin permits, and await a "green deck."

NOTE: If an emergency condition exists that requires an immediate recovery, the pilot should flash the aircraft position lights, anticollision lights, or the landing light to expedite the "green deck." If no other emergency exists that requires an immediate recovery, the pilot should turn the aircraft position lights on steady.

(d) The "green deck" status light will be turned on to notify the flight that the ship is prepared to recover the aircraft. When the aircraft are on short final, the deck status light should be extinguished.

b. ***Lost Radar Contact.*** If the flight is to be conducted under positive radar contact and contact is lost, the controller should notify the flight immediately. The controller will give the flight a vector from the last known position of the flight that will bring the flight back toward the ship.

APPENDIX E

STANDING OPERATING PROCEDURE FOR OVERWATER OPERATIONS

The appendix provides a sample format for an overwater SOP. Units may use this sample when developing their own overwater SOPs.

1. Purpose. The information and procedures in this SOP will enable all assigned aviators to master the techniques needed to conduct operations in an overwater environment. Overwater operations are those operations conducted outside the gliding distance of the shoreline.

2. Scope. These procedures apply to all aviators assigned to _______________ (unit) and to all individuals acting as crew members in aircraft assigned to _______________ (unit).

3. Overwater Flying Techniques and Procedures.

a. All flights operating in an overwater environment will have, as a minimum, the following crew mixes:

(1) Aircraft must have two pilots that meet the criteria described in the appropriate aircrew training manual. Single-pilot operations are not authorized for training in overwater operations.

(2) At least two aircraft are required for overwater operations. This provides for mutual support in case one aircraft has an emergency that requires extraction from the water. Special considerations are required for single-engine aircraft (OH-58(I)) and for dedicated SAR aircraft. UH-60 aircraft may perform single aircraft overwater flight when specific requirements are briefed during the mission briefing/OPORD.

b. To enhance mission safety, many situations will require varying altitudes. Unless stated differently in the operations order, the standard flight altitude for overwater day and/or NVG training is 50 to 80 feet above the water.

c. The base airspeed for operations conducted in an overwater environment is 80 KIAS for the OH-58D(I) and 100 KIAS for the AH-64 and UH-60.

d. The standard formation for overwater flight is echelon right or left. This will be briefed before multiship operations are conducted. The normal rotor separation shall be ten rotor diameters. Because of the lack of visual cues to determine rates of closure, the trail formation is the most difficult formation to fly. Therefore, straight trail formations should be avoided. During periods of reduced visibility and illumination, flying closer than the required five to eight rotor diameters may become necessary.

(1) In the assessment of the tactical situation, the tighter the formation the easier it becomes for threat radar to acquire the flight. Tight formations flown in coastal areas also reduce aircraft maneuvering area. This becomes hazardous when flocks of waterfowl are encountered.

(2) During overwater flight, aircraft should not be overstacked or understacked. This is especially critical in turns at low altitudes. Overwater formation flying requires a great deal of discipline. Pilots should take care that they do not fall far behind the aircraft to the front. Falling back may result in the visual loss of the flight.

e. The minimum weather for training is 1,000-foot ceilings and 3 SM visible horizon in two quadrants of the horizon. This is one of the most critical aspects of the overwater environment because of the rapid changes that may occur. In addition, signs of impending bad weather over water are not as obvious as they are over land. Weather forecasts for overwater operations will contain the following information:

(1) The sea state to include height, direction, current, fetch, and distance between the swells.

(2) The temperature of the sea.

(3) The availability of any overwater remote stations.

(4) The temperature and dewpoint spread. (This is important in determining the formation of sea fog.)

(5) The high and low tides.

f. Effects of the weather will determine how most tasks are conducted. This is demonstrated by the effects of wind on the sea state. The wind will increase the size of swells to the point that ship operations may become impossible because of pitch and roll angles created by the swells. The pitch and roll limitations for ship operations are ten degrees roll and five degrees pitch. If, however, the wind is calm, the surface of the water becomes smooth, resulting in the loss of height and motion perception. All weather factors encountered over land have different effects over water. Planners must carefully consider the effects of weather on overwater operations.

g. Aircraft used in an overwater environment must have the following equipment installed and operational:

(1) Radar altimeter (properly calibrated).

(2) Rotor brake (AH-64, for shipboard operations).

(3) Extraction ladder or caving ladder. (OH-58, UH-60)

(4) One raft per crew member in crew station (AH-64, OH-58D(I)).

(5) One overwater kit per crew member.

(6) Two seven-man rafts (UH-60).

NOTE: Crew doors will be removed on OH-58D(I) and UH-60 aircraft.

h. When the approach direction to a target in an overwater environment is being planned, the noise of the helicopters must be masked. To do this, the flight must use wind direction to reduce maneuvering and avoid easily detectable power changes. To avoid excess noise, the flight should approach the target into the wind. This also will help the flight conduct approaches. Moon angle, illumination, and background lights are all considerations when a target is being approached. Canopy reflections and moon reflection from the water help to identify aircraft from long distances.

i. Depending on the surface of the water and weather conditions, pilots may find the transition from overland altitudes to overwater altitudes difficult. This transition is critical in overwater operations and must be planned carefully. At the shoreline, the pilot may plan so that turns are parallel the coast line and the coastline can be kept in sight while making the descent to overwater altitudes. This enables the pilot of the lead aircraft to gain visual contact with the water, thus making a smooth transition.

j. When descending from altitude to overwater flight profiles, pilots must ensure that the descent does not exceed three hundred feet per minute. This will help the pilots avoid miscalculating the descent and crashing into the water. During descents in this environment, crew coordination is extremely important. The pilot should concentrate on flying the aircraft. The copilot should read off the rate of descent and the altitude from the radar altimeter constantly. The low light of the radar altimeter is very important in warning the pilot of low altitudes. This light will be set 10 feet below the altitude that the flight intends to hold. When setting the low light of the radar altimeter, the person doing the setting should announce the setting he makes. This will ensure that both crew members know at what point the light will come on.

k. In an overwater environment, few hazards with vertical development exist. However, the types of hazards associated with this environment are flocks of birds, channel buoys, boats, and so forth. Planners must be aware of these hazards when planning the en route portion of the mission. Maritime charts show the location of channel buoys and the type of lighting associated with these buoys. This lighting may be steady or it may flash in a set sequence. Planners also should avoid shipping lanes when analyzing the maritime charts.

l. Navigation is performed using vectoring, on-board navigation systems, and dead reckoning. The following list of techniques enable navigators to perform navigational tasks when using dead reckoning:

(1) Navigating to a specific point along a coastal area. Navigating to a specific point along a coastal area may be difficult until the flight is directly over the location. This is because of the lack of vertical development of coastal areas. This fact, along with the navigator's inability to determine intermediate times en route, makes this type of navigation very difficult. To ensure that the flight arrives at the first ACP on shore, planners should offset the route to one side of the ACP. This intentional deviation assures the navigator of his location in respect to the ACP. The navigator always will know which way he must turn to reach the ACP. If the route is planned directly at the ACP on shore, the flight may end up either to the left side or right side of the ACP. The navigator must then determine which way to turn to fly to the ACP. The offset method eliminates the need for guesswork and replaces it with a solid course of action. To use the offset method, planners must consider the winds, magnetic variation, and so on. Once these factors have been considered, planners can plot the course to offset in a direction that minimizes or compliments the effects of the wind.

(2) Navigating from onshore to offshore. The flight must find the closest reference point or ACP to the target. For example, a lighthouse or channel marker may be used as the last ACP before the time/heading mode is entered. This technique reduces navigation errors. In any type of overwater navigation, some type of reference points must be used along the course line drawn on the map. At the preference of the navigators, these marks may be mile tick marks or time tick marks.

m. For crew member survivability during overwater operations, the doors of the aircraft will be removed or opened. This allows for immediate egress in case the aircraft enters the water. The PC of each aircraft will brief all crew members on procedures to be used if the aircraft enters the water. The procedures learned in the 9D5 Dunker serve as a guide for egression from the aircraft. Personal floatation devices must be worn at all times when outside the gliding distance from shore. Crew members will ensure that all personal items are secure and that they will not catch on any part of the aircraft and hinder egression. Before conducting overwater operations, the crew should practice the egress procedures to be used during black-out conditions. LPUs should not be inflated until crew members are well clear of the aircraft.

n. The immersion suit is an integral part of the aviator's ALSE. The immersion suit helps prevent hypothermia in case the crew enters the water. It also helps provide crew members with flotation. Immersion suits will be worn during overwater operations when the water temperature is below 60 degrees Fahrenheit. When the water temperature is between 61 and 70 degrees Fahrenheit, the unit commander or his designated representative may waive the wearing of the immersion suit. When the water temperature is above 71 degrees Fahrenheit, the wearing of the immersion suit is at the discretion of the individual pilot.

o. During overwater flight, one of the most critical areas of concern is the texture of the water surface. The texture of the water surface will vary from extremely rough during high winds to a mirror-like surface when the winds are calm. The rougher of the water, the easier it is to judge altitudes above the water. A smooth water surface may induce many illusions because pilots cannot sense motion or determine the height of the aircraft above the water. When the surface of the water is smooth, the aircraft may have to be flown higher. Radar altimeter cross check also must be completed more frequently. Smooth surfaces may induce the sensation of being too high above the water. The pilot may react by placing the aircraft into a descent toward the water. Spacial disorientation also may occur over smooth water because of the reflection of the stars in the water. To prevent vertigo when flying over smooth surfaces, pilots must trust the radar altimeter and their other flight instruments.

p. The overwater operational procedures checklist (premission planning) is as follows:

(1) Maps (including maritime).

(2) Weather (including sea data).

(3) Aircraft equipment (including the following):

(a) Aircraft cockpit lighting.

(b) Radar altimeters.

(c) Operable parking brakes.

(d) Operable tail wheel lock pins.

(e) Blade folding equipment (if available).

(4) Navigation equipment/vector platform (VOR/NDB).

(5) ALSE (briefed in detail in the OPORD).

TAB A. Overwater Extraction Procedures and CSAR/Downed Aircraft Procedures to the Overwater SOP

1. Purpose. To provide and describe the basic techniques of overwater extraction of a downed aircrew.

2. Scope. Although some techniques described in this SOP may be used over land, these procedures apply to overwater extraction only.

3. General. Current policy requires that UH, AH, and OH aircraft operating overwater will not be deployed single ship or single pilot. This policy must be strictly adhered to and only in an emergency may it be deviated from. If overwater operations are attempted single pilot, extraction procedures can become extremely dangerous, if not impossible.

4. Aircraft Crew and Configuration.

a. Each UH-60 crew will consist of two pilots and two crew chiefs. The SAR aircraft will have SAR swimmers and medical personnel on board.

b. Each UH-60 will have two ladders (Jacobs or caving) on board--one for each cargo door. Chemical lights must be available so they can be taped to the bottom of each ladder if the ladders are to be used during darkness. The PC also should ensure there are enough chemical lights are available to be used for visual references during an extraction. Each UH-60 will carry two seven-man rafts, one overwater survival kit, one large first aid kit, and two saltwater-activated flares. Three aircraft first aid kits also will be carried at all times.

c. If possible, SAR aircraft will be equipped with a right-side external hydraulic hoist capable of lifting 600 pounds. A jungle penetrator with a flotation collar and a horse collar also will be carried to use with the hoist.

5. Crew Responsibilities.

a. Except during emergency situations, each aircraft that performs extraction operations must be dual-pilot.

b. PCs have overall responsibility for assessing the situation and supervising the operation regardless of the seat they occupy. They also are responsible for the installation and preflight of the extraction equipment.

c. The pilot in the left seat is responsible for flying the aircraft .

d. The pilot in the right seat is responsible for deploying the caving ladder and directing and guiding the helicopter by giving verbal commands to the pilot in the left seat. He also monitors the altitude, power, and hookup of the downed aviator.

e. The crew also will determine the maximum allowable gross weight of the aircraft and determine the number of personnel that can be extracted at one time.

6. Aircrew SAR- and CSAR-Related Personnel Requirements.

a. All aircrew members will have the equipment listed in the unit SOP.

b. Downed aircraft personnel should use the strobe light to help SAR aircraft locate them from greater distances. After the SAR aircraft arrives overhead, the chemical lights should be switched on. Downed personnel also will turn on their survival radios and leave them on the prebriefed channel.

7. Preflight.

a. The caving ladder, which is used for extraction, is a 30-foot, high-strength aluminum ladder. Before attaching the ladder to the aircraft, the PC will visually inspect the ladder for obvious defects in support wires and rungs. When possible, the ladder will be attached the right forward clevis ring. The PC will ensure that the D-ring is attached to the support wire and not to the rung. This acts as a safety if the grummel hooks should break. A new red chemical light should be taken out of its protective wrapper and broken before the ladder is deployed. When preparations have been completed, the ladder is rolled out completely and rerolled. The ladder is stored in its protective canvas bag and placed where it is easily accessible to the right-seat pilot. The right-seat pilot should rehearse the procedures for deploying the caving ladder.

b. The system will be checked thoroughly for serviceability, operation, and security.

c. Whenever possible, extra flotation gear, such as overwater survival kits and one-man rafts, may be carried for the downed crew.

d. All crew members must be briefed thoroughly and understand all aspects of the extraction procedures.

8. SAR Information. To ensure success, every SAR and CSAR operation requires certain information. A common air-to-air frequency should be established among the flight to help the SAR aircraft with the recovery. As a minimum, aircraft at the downed aircraft site should transmit to the SAR aircraft the following information:

a. Call sign, type of aircraft, and the number of personnel involved.

b. Location or last known position (UTM/latitude-longitude, magnetic heading, and distance from the CP).

c. Condition of personnel, if known.

d. Amount of time the flight can remain on station to assist the SAR aircraft.

e. Weather and enemy situation.

9. Overwater Extraction Procedures.

a. Operational and illumination requirements will dictate whether downed aircrew will be recovered quickly or wait for another method of SAR such as small boat or ship, if any are in the area. The AMC or the SAR crew generally makes this decision. All attempts will be made to extract downed crews in a training environment.

b. The SAR aircraft that arrives on station need help to determine the exact location of the survivors. If the situation permits, survivors should give vectors to the SAR aircraft. The pilot at the controls begins a right or left turn and flies an oval pattern at a comfortable altitude and airspeed. The pilot not on the controls monitors the survivors, and the crew chiefs prepare to deploy the chemical lights. As soon as the survivors are sighted, light markers will be deployed to aid in spatial orientation and pickup.

c. On the initial pass of the SAR aircraft, survivors should indicate if they are able to climb the ladder by waving one of their arms over their head. At night, they should wave a chemical light. If survivors remain still in the water, this indicates that they are either unconscious or unable to climb the ladder. The aircrew will mark the location of the survivors with several chemical lights tied together. If the survivors appear to need help, the SAR aircraft will fly a pattern to pass over the survivors. This pattern will be flown into the wind at low altitude and airspeed (10 feet and 10 knots). The SAR aircraft continues into the wind to set up an oval traffic pattern. If the survivors are not moving or do not indicate that they can climb the ladder, rescue personnel will be alerted to prepare to helocast into the water with their extraction equipment (horse collar, stokes litter, and so on).

d. The pilot on the controls will fly the pattern to pass over the survivors and rescue personnel. The pattern will be into the wind at approximately 30 feet and 40 KIAS. As the survivors pass under the nose of the aircraft, a set of chemical lights will be deployed out of each side of the aircraft. This will be done three more times at one-to-two second intervals. Deployment of the chemical lights in this manner creates a "runway" to line up on and use for visual reference during the extraction. The pilot on the controls will continue around the pattern one more time. On final, he will begin a decelerating approach to arrive just short of the survivors at a slow hover (1 to 30 feet depending on sea state and visual cues available). The aircraft will continue to move forward using the survivors and chemical lights as a line-up reference.

e. At a slow hover short of the pickup point, the ladder is removed from the bag and deployed. The ladder will be dropped outside the skid or wheel strut. The ladder may not fully unravel after it is dropped. If this occurs, the ladder should be grabbed and jerked up on abruptly. This action will deploy the ladder fully.

(1) The ladder should be deployed about 100 feet from the survivors. The 30-foot ladder will enter the water at a 27-foot radar altitude. The pilot should continue hovering slowly toward the survivors at a 20- to 25-foot radar altitude. If possible, the approach should be made into the wind. To maintain a stabilized hover, the pilot will use all available references such as lights on the horizon, trees, or boats. The tendency is to lose altitude as the survivors are approached. At a 15-foot radar altitude or less, the rotor systems will recirculate salt spray which will partially obscure the windshield. If no references are available, the pilot should deploy chemical lights to maintain a stable hover as shown below.

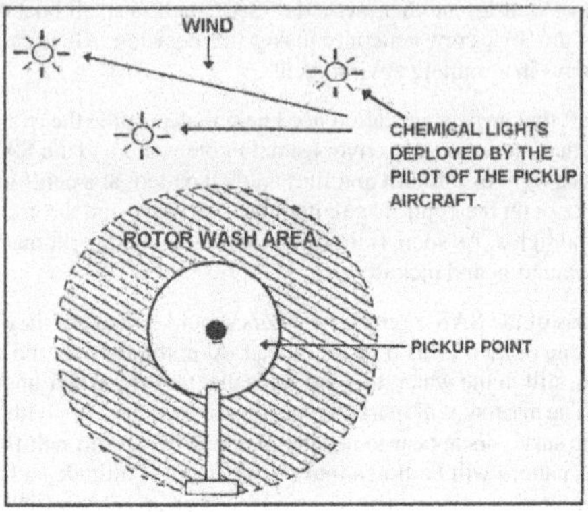

(2) If the survivor only hooks up, the pilot will advise the left-seat pilot to pick up slowly as the right-seat pilot monitors the survivors. The survivors should be lifted vertically out of the water to avoid severe pendular actions.

(3) If the survivor can climb the ladder, the right-seat pilot directs the aircraft down as the survivor starts his ascent. If done properly, survivors climbing the ladder should be no more than 2 feet above the water at any time. This prevents injuries should the survivor fall off the ladder. If the sea state permits, a good techniques is to hover as close to the water as possible and allow the survivor to climb into the aircraft. Caving ladders are limited to 500 pounds (two personnel) and Jacobs ladders are limited to 750 pounds (three personnel).

(4) If taking off with survivors hooked onto the ladder becomes necessary, the airspeed should be no more than 40 knots. The altitude should be high enough to keep the survivor at least 10 feet from the water. Pilots should avoid higher altitudes. At this altitude, the survivor will not be hurt if he falls from the ladder. When the survivor is dropped, the aircraft may need to return for the remaining survivors. If so, the ladder should be pulled back inside or the aircraft flown at less than 30 knots of airspeed. This will keep the ladder from getting caught in the tail rotor.

g. The downed aircrew will take the actions described below.

(1) Once the rescue aircraft is clear of the downed aircraft, check to see if anyone else has surfaced nearby. If someone has surfaced, link up if possible. If linkup is not possible or if you are the only survivor, turn on every available light.

(2) Take out and deploy your five chemical lights. Save the fifth light to signal your status (injured or uninjured). When the extraction aircraft is over you and if you can climb the ladder, wave the fifth chemical light back and forth above your head. As shown below, this signal indicates that you are unhurt and that you will try to climb the ladder.

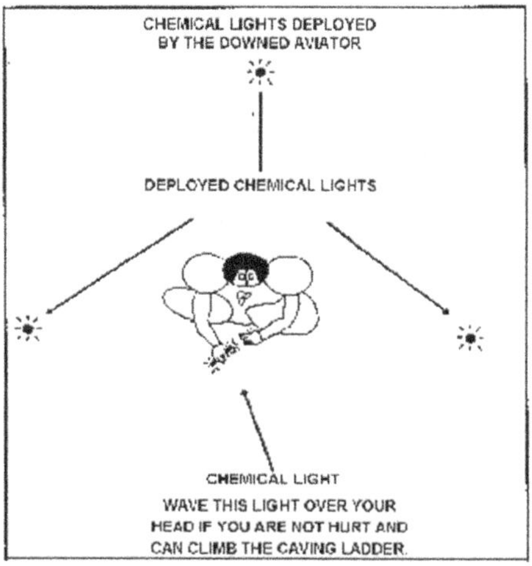

(3) To keep from being separated during recovery, survivors should hold onto each other until they are safely on the hoist or ladder. Before grabbing hold of the ladder, allow the hoist and the ladder to touch the water to dissipate any static electricity.

(a) If the penetrator is used, the survivors must unhook the seat release and position themselves on the seat. Then they will attach the safety belt, make sure that no loose cable is wrapped around them, and give a thumbs-up signal to be lifted. (At night, the chemical light will be held high over the head.) If ladders are used, survivors should climb up into the aircraft. If this is not possible, they should hook into the ladder using a D ring. The D ring will be hooked into the cable instead of the step.

(b) Before climbing the ladder, unhook the coupling clips on the waist portion of the LPU. This enables you to get close enough to the ladder to climb it. The best technique to use in climbing a ladder is to climb the side of the ladder using your feet to push yourself up and your arms to maintain stability on the ladder. If you cannot climb the ladder or you get tired on the way up, use your D ring to hook onto the ladder. The D ring should be hooked to the side support wire of the ladder in case the rung breaks. Do not hook the D ring onto the rungs of the ladder. No more than two individuals should be on the ladder at one time.

Figure E-1. Suggested format for an overwater SOP

APPENDIX F

FLIGHT DECK CLOTHING AND DUTIES

The Navy and Coast Guard use standardized clothing color schemes for flight deck personnel. This allows crew members to quickly identify personnel by duty on a busy flight deck. Table F-1 shows these standardized color schemes.

Table F-1. Standardized clothing color schemes

PERSONNEL	HELMET[1]	JERSEY	SYMBOLS[2]
Aircraft Handling Crew and Chockmen	Blue	Blue	Crew Number
Aircraft Handling Officers, CPO, LPO	Yellow	Yellow	Billet Title
Elevator Operators	White	Blue	E
LSE (Crew Directors)	Yellow	Yellow	Crew Number
Maintenance Crews	Green	Green	Black Stripe and Squadron Designator
Medical	White	White	Red Cross
Messengers and Telephone Talkers	White	Blue	T
Photographers	Green	Green	P
Plane Captains	Brown	Brown	Squadron Designator
Ordnance	Red	Red	Black Stripe and Squadron Designator/ ship's billet title
Crash and Salvage Crews	Red	Red	Crash/Salvage
Tractor Driver	Blue	Blue	Tractor
Maintenance Crews	Green	Green	Black Stripe broken by abbreviation of specialty (ie P/P (Power Plants))
Aviation Fuel Crew	Purple	Purple	F
Aviation Fuel Officer	Purple	Purple	Fuel Officer
Combat Cargo	White	White	Combat Cargo
Safety Observer	White	White	Green Cross

NOTES:

1. The helmet referred to is a combination cranial helmet.

2. USCG flight deck clothing does not include symbols.

F-1. GENERAL

a. The vest type life preserver (US Navy, Mk-1) is designed for prolonged use while the wearer is engaged in flight deck activity. It is available in colors identical to those in Table F-1.

b. Combination cranial helmets for the personnel listed below shall be marked with three vertical reflective international orange stripes. The stripes will be one inch wide, evenly spaced, and placed on top of white reflective tape.

- All officers.
- Flight and hangar deck chief petty officer and leading petty officer.
- Crash and salvage chief petty officer and leading petty officer.
- EOD team members.
- Squadron's ordnance officer.
- Ship's air gunner.

c. Helmets for all other personnel shall be marked with a 6-inch square (or equivalent) of white reflective tape on the back shell. It will have a 3-inch by 6-inch piece (or equivalent) of white reflective tape on the front shell.

d. At night, the ordnance arming/safety supervisor will carry two red standard wands. These wands will be banded with two 3/4-inch bands that are equally spaced on the cones.

F-2. FOREIGN OBJECT DAMAGE WALKDOWN. These walkdowns are conducted before, between, and after flight operations. Personnel who are not standing watch will form a line across the width of the flight deck and walk slowly from bow to stern in search of solid particles, bolts, screws, and so on that could be ingested into the intake of an aircraft engine and cause severe damage or engine failure. FOD is a major safety concern on all flight and hangar decks.

F-3. DUTIES

a. ***Yellow Jerseys.***

(1) ***Aircraft handling officer.*** The ACHO or "handler" controls all movement of aircraft on the flight and hangar decks from flight deck control. He also maintains a running maintenance status of every aircraft on board and coordinates with weapons, hangar deck control, intermediate maintenance, supply, and air operations personnel. Because of limited space on the flight deck and hangar deck, he coordinates both space allocation between other departments and the use of the aircraft elevators.

(2) ***Flight deck officer.*** The FDO directs, plans, and oversees the parking and security of all aircraft and mobile fire-fighting and ground support equipment on the flight deck. He also supervises a large division that includes all aircraft directors, plane handlers, tractor drivers, elevator operators, and crash and salvage personnel as well as his administrative staff.

(3) ***Landing signal enlisted/aircraft director.*** Under the supervision of the air officer, the

LSE is responsible for visual signals to the helicopter pilots which help them make safe takeoffs and/or approaches and landings. He directs the pilot to the desired deck spot and ensures that general safety conditions are maintained within the immediate helicopter landing zone. He ensures that, on signal, helicopters are started, engaged, launched, recovered, and shut down safely. The LSE also ensures that all tie-downs are removed before lift-off and secured properly after landing. Except for the wave-off and hold signals, which are mandatory, the LSE's signals are advisory in nature. As crew director, he is in charge of aircraft handling crews and ensures that aircraft movement on deck is conducted safely and according to NATOPS procedures.

b. ***Red Jerseys.***

(1) ***Crash and salvage.*** The flight deck "fire department" fights aircraft fires and rescues personnel on the flight deck. They operate all mobile fire-fighting and crash/salvage equipment.

(2) ***Ordnance officer.*** The ordnance officer is responsible for the safe movement, handling, and loading of aircraft ordnance. His jersey has a black stripe and the words"Ordnance Officer" printed on the front and on the back.

(3) ***Explosive ordnance disposal officer and crew.*** These personnel dispose of, disarm, and neutralize defective ordnance. Their jerseys have the letters "EOD" printed on the front and on the back.

(4) ***Ordnance handlers.*** The ordnance handlers or "B-B stackers" move, load, and unload ordnance on the aircraft. Their jerseys have black stripes and their squadron designator or ship billet printed on the front and on the back.

c. ***White Jerseys.***

(1) ***Safety officer and crew.*** The safety officer and crew are responsible for the overall safety of flight operations. They make sure that all flight deck activities are conducted according to established safety procedures.

(2) ***Combat cargo officer.*** The CCO coordinates and executes the loading, unloading, and movement of all air cargo and passengers. The back of his jersey has the words "Combat Cargo" printed on it.

(3) ***Medical.*** Medical personnel provide immediate medical assistance and treatment to any flight deck personnel casualties. Their jerseys have a large red cross on the front and on the back.

d. ***Purple Jerseys.*** Known as "grapes" because of the color of their jerseys, these personnel fuel and defuel aircraft using fuel stations located on the flight and hangar decks.

e. ***Blue Jerseys.***

(1) ***Aircraft handling crew.*** The "blue shirts" are responsible for handling and tying down all aircraft with chocks and chains. They also operate the handling equipment, including tractors and electrical power units on the flight deck.

(2) ***Elevator operators.*** The EOs operate the ship's aircraft elevators, which move aircraft to and from the flight deck and hangar deck. They wear blue shirts and white helmets.

f. ***Green Jerseys.***

(1) ***Squadron maintenance crew.*** The squadron maintenance crew maintains the Navy and Marine Corps aircraft on the ship. Their jerseys are marked with their squadron designator and a black stripe on the front and on the back.

(2) ***Ground support equipment maintenance crew.*** These personnel maintain the ground support equipment that is assigned to the flight or hangar deck.

(3) ***Photographers.*** These personnel photograph and videotape flight operations for documentation and media requests.

g. ***Brown Jerseys.*** Plane captains, both Navy and Marine, wear brown jerseys. They ensure that the aircraft are inspected properly and serviced before and after each flight. They also supervise ground starting procedures and are responsible for the cleanliness and general condition of their aircraft. Their jerseys are marked with their squadron designator on the front and on the back.

APPENDIX G

HELICOPTER/SHIP INTERFACE

This appendix shows the compatibility between Army helicopters and Navy and Coast Guard ships.

Section I. USN/USCG Ship and Army Helicopter Compatibility

G-1. SHIP/HELICOPTER OPERATIONS

The data shown in the tables and figures in this appendix are extracted from NWP 3-04.1M and are provided for information only. If differences exist between NWP 3-04.1M and the information in this manual, NWP 3-04.1M takes precedence.

a. ***Explanation of Tables.*** Table G-1 (page G-3) gives the definition of the acronyms referred to in Tables G-2 (page G-4) and G-3 (page G-6). Tables G-2 and G-3 show ship and helicopter combinations suitable for takeoff, landing, and vertical replenishment.

(1) Helicopter rotor diameter, fuselage configurations, landing gear arrangement, maximum gross weight, ship structures, and flight deck obstructions were considered to obtain the data in these tables. The tables show combinations of ships and aircraft that are suitable for safe operations during clear weather conditions without servicing facilities.

(2) The combinations do not constitute authority to operate nor do they imply certification of various facilities for the helicopters noted. Ships are listed based on their designed facilities. Certification of each individual facility must be verified before actual operations at that facility. Preferably, this will be done during the presail conference.

b. ***Table Annotations.***

(1) In all cases, the operating helicopter is restricted to landing with the fuselage parallel to the landing lineup line and the forward landing gear or skid support within the touchdown circle inner edge.

(2) In Table G-2 under the column, "VERTREP TYPE," Type 1, Type 2, and Special Type 2 operations are defined as follows:

(a) ***Type 1 operations.*** The helicopter must hover with all rotor hubs above the segmented lineup line.

(b) ***Type 2 operations.*** The helicopter must hover with all rotor hubs above or aft of the "T" lineup line.

(c) ***Special type 2 operations.*** In general, pilots of helicopters that have a rotor diameter of less than 55 feet must follow the guidance for VERTREP Type 2 operations. Helicopters that have a rotor diameter of more than 55 feet must be hovered with all rotor hubs above or aft of the "T-ball" lineup lines.

(3) In Table G-2 under the column, "VERTREP HELICOPTER," Class 4 and Class 5 VERTREP operations are defined as follows:

(a) Class 4 VERTREP operations require that the helicopter be hovered at a minimum altitude in excess of 5 feet above the operating deck.

(b) Class 5 VERTREP operations require that the helicopter be hovered at a minimum altitude in excess of 15 feet above the operating deck.

(4) In Table G-3 under the column "OPERATING LEVEL," the letter "I" denotes a day/night IMC capability; the letters "II" denote a day/night VMC capability; and the letters "III" denote a day VMC capability.

(5) In Table G-3, the dual-circle, air-capable ship matrix addresses all helicopters that may be operated simultaneously on the same deck. Clearances for any particular helicopter landing at the aft circle are assured only when a parked helicopter or any other object remains forward of the safe park line.

WARNING

Some Army helicopters have not been tested in the electromagnetic environment of all classes of ships. When operations are being conducted with Navy ships, consideration must be given to potential radiation hazards, electromagnetic interference, and electronic vulnerability effects.

G-2. SHIP DESIGNATIONS

a. ***Definition of Acronyms.*** Table G-1 lists the definitions of ship acronyms that are referred to in Tables G-2 and G-3.

Table G-1. Acronym definitions

AD	destroyer tenders
AE	ammunition ships
AFS	combat store ships
AGF	command ships
AO	oilers
AOE	fast combat support ships
AOR	replenishment oilers
AR	repair ships
ARS	salvage ships
AS	submarine tenders
ASR	submarine rescue ships
ATF	fleet ocean tugs
ATS	salvage tugs

CG	guided missile cruisers
CV	aircraft carriers
DD	destroyers
DDG	guided missile destroyers
FF	frigates
FFG	guided missile frigates
IX	helicopter landing trainer
LCC	amphibious command ships
LHA/LHD	amphibious assault ships
LKA	amphibious cargo ships
LPD	amphibious transport dock ships
LPH	amphibious assault ships
LSD	dock landing ships
LST	tank landing ships
T	military sealift command ships
W	Coast Guard cutters

b. ***Explanation of Suffixes and Prefixes.***

(1) The suffix "N" denotes a nuclear powered vessel.

(2) The prefix "T" refers to a tactical ship. On a tactical ship, the crew consists of military sailors. A nontactical ship is one that belongs to the Military Sealift Command. These ships have civilian crews and a military captain.

c. ***Listing of Ships.*** All ships are not individually listed in Tables G-2 and G-3. Rather, they are listed by class. For example, DDG 993 (USS Kidd) represents four ships: DDGs 993, 994, 995, and 996. To find a specific ship, refer to NAEC-ENG-7576, the Navy's catalog of aviation-capable ships. Once the specific ship is identified in the resume manual, its specific capabilities can be found in this appendix.

Table G-2. Single-circle air-capable ships

				LANDING HELICOPTER						VERTREP HELICOPTER	
SHIP	AREA	SS OR CLASS	LEVEL/ CLASS	UH-1	OH-6	OH-58D	UH-60	CH-47	VERTREP TYPE	CH-47	UH-60
AD 38	AFT	S	II, 2A	X	X	X			SP2	4	4
AD 41	AFT	C	II, 2A	X	X		X	X	SP2	4	4

AE 27	AFT	C	I, 2A	X	X	X	X		SP2	4	4
AGF 3	AFT	S	I, 2A	X	X	X	X		SP2	4	4
AO 177	AFT	C	II, 3	X	X	X			SP2	4	4
AOE 1	AFT	C	I, 2A	X	X	X	X		SP2	4	4
AOE 6	AFT	C	I, 2A	X	X	X	X		SP2	4	4
AOR 6	AFT	S	I, 2A	X	X	X	X		SP2	4	4
CG 47	AFT	C	I, 2A	X	X	X			2	4	4
	MAIN		I, 5						1	5	5
	FWD		III, 4						1	4	4
CGN 36	AFT	C	I, 4	X			X	X	2	4	4
	FWD		III, 4						1	4	4
DD 963	AFT	C	I, 2A	X	X	X			2	4	4
	MAIN		I, 4						1	4	4
	FWD		III, 4						1	4	4
DDG 51	AFT	C	I, 4	X	X	X			2	4	4
	FWD		I, 4						1	4	4
DDG 993	AFT	C	I, 2A	X	X	X			2	4	4
	MAIN		I, 4						1	4	4
	FWD		III, 4						1	4	4
FFG 7	AFT	C	I, 2A		X	X			SP2	4	4
	FWD		III, 5						1	5	5
IX 514	MAIN	S	II, 3	X	X	X	X		2	4	4
LCC 19	AFT	C	I, 2A	X	X	X	X	X	2	4	4
LSD 36	AFT	C	II, 2A	X	X	X	X		2	4	4
T-AE 26	AFT	S	I, 2A	X	X	X	X		SP2	4	4
T-AFS 1	AFT	C	I, 2A	X	X	X	X		SP2	4	4
T-AFS 8	AFT	C	I, 2A	X	X	X			SP2	4	4
T-AH 19	MAIN	C	I, 2A	X	X	X	X	X	3	4	4
T-AK 3000	AFT	C	II, 3	X	X	X	X		2	4	4
T-AK 3005	AFT	C	II, 3	X	X	X	X		1	4	4
T-AK 3008	AFT	C	II, 3	X	X	X			1	4	4
T-AO 188	AFT	C	II, 3	X	X	X	X		2	4	4
WHEC 715	AFT	C	I, 2A	X	X	X			2	4/5	4/5
WMEC 901	AFT	C	I, 2A	X	X	X	X		2	4	4

Table G-3. Dual-circle air-capable ships

SHIP CLASS	OPERATING LEVEL	HELICOPTER LANDING IN: ①																									
		FORWARD LANDING CIRCLE													AFT LANDING CIRCLE												
		H1	H2	H3	H6	H46	H47	H53	H53E	H54	H56	H60	H64	H65	H1	H2	H3	H6	H46	H47	H53	H53E	H54	H56	H60	H64	H65
LPD 1	1	X	X	X	X	X	X	X		X	X	X	X	X	X	X	X	X	6	X	6		6	X	X		x
LPD 4 ②	1	X	X	X	X	X	X	X		X	X	X	X	X	X	X	X	X	6	X	6		6	X	X	X	X
LSD 41	H	X	X	X	X	X		X	X	X	X	X	X	X	X	X	X	X	X	X	X	X	X	X	X	X	X

Notes:

1. X: Full-circle landing
   3: Less than full-circle landing restriction
   4: Low-hover Class 4 operation
   5: High-hover Class 5 operation
   Blank: Operational incompatibility
2. Includes AGF II

Table G-3. Dual-circle air-capable ships (continued)

DUAL-CIRCLE AIR-CAPABLE SHIP/US MILITARY HELICOPTER INTEROPERABILITY													
HELO PARKED IN AFT CIRCLE	HELO LANDING IN FORWARD CIRCLE ①												
	H 1	H 2	H 3	H 6	H 46	H 47	H 53	H 53 E	H 54	H 58	H 60	H 64	H 65
H-1	BC	BC	BC	ABC	BC	B				ABC	BC	BC	BC
H-2	ABC	ABC	BC	ABC	BC	B				ABC	BC	ABC	ABC
H-3	BC	BC		BC	C					BC		BC	BC
H-6	ABC	ABC	BC	ABC	ABC	AB	BC		BC	ABC	ABC	ABC	ABC
H-46	BC	BC		ABC	BC					BC	BC	BC	BC
H-47	BC	BC		ABC	BC					BC	BC	BC	BC
H-53		C		BC						BC			
H-53E				C						C			
H-54		C		BC						BC			
H-58	ABC	ABC	BC	ABC	ABC	B				ABC	ABC	ABC	ABC
H-60	BC	BC		ABC	BC					ABC	BC	BC	BC
H-64	BC	BC	BC	BC	BC	B				BC	BC	BC	BC
H-65	BC	ABC	BC	ABC	BC	B				ABC	BC	BC	ABC

Notes:

1. A: LPD 1 Class
   B: LPD 4 Class
   C: LSD 41 Class

Section II. Army Helicopter Wind Envelopes

G-3. AIRCRAFT WIND LIMITATIONS

a. The safe launch and recovery wind limitations are presented in NWP 3-04.1M and COMDTINST M3710.2-series. These publications can be obtained using the procedures outlined in Chapter 1.

(1) Figures G-1 (page G-10) and G-2 (page G-11) are to be used when specific flight limit envelopes are not available. Procedures should be agreed upon before starting the approach. The envelopes are for a normal approach to the spot and the helicopter aligned the with ship's centerline at touchdown. The two wind envelopes present wind-over-deck limitations relative to the ship and represent safe operating conditions as tested to date.

(2) Heavy lines on the envelopes denote where specific problems were encountered during testing such as approaching a flight control limit or excessive turbulence. Ship pitch and roll limitations are left to the judgment of the ship's commanding officer and embarked aviation unit commander except where listed on the appropriate envelopes.

WARNING

Considerable differences may exist between the flight deck winds and those measured by bridge level anemometers. However, aircraft wind limitations contained in this appendix and shipboard operating bulletins are based on winds measured by the windward bridge-level anemometer. When operations are being conducted near the outer wind limits, the probability of damage increases sharply when wind gusts exceed 10 knots. Also, the maximum safe wind in conjunction with excessive ship pitch and/or roll can make flight operations unacceptably hazardous. Common sources of turbulence are stack gasses and wash, ship superstructures, deck protrusions, and rotor wash or jet blast caused by adjacent aircraft taking off and landing.

b. The Naval Air Warfare Division, Patuxent River, Maryland, conducts dynamic interface testing of all helicopter and ship class combinations to develop all aspects of shipboard helicopter dynamic operational compatibility. DI testing investigates the effects of ship airwake, ship motion, and ship lighting and marking. It also investigates the effects of ship and helicopter operations. The Army Shipboard Testing Office at NAWCAD, DSN 342-1342 or 342-1336, conducts DI testing of Army aircraft.

(1) The significant result of DI testing is the development of operational launch and recovery, engage or disengage, and helicopter in-flight refueling envelopes. Each of these depicts the wind, speed, direction, and ship motion conditions conducive to producing consistently safe shipboard operations. DI certifications of each ship and helicopter combination are required before any shipboard flight operations are conducted that are beyond the bounds of the DI general envelope.

(2) Aircraft that have not undergone DI testing or that do not have a DI-certified envelope are restricted to the use of the general launch and recovery wind limitation charts for the appropriate class ship. Comments or questions about the wind envelopes should be addressed to:

Commander
Naval Air Systems Command
(Code PMA 251, Aircraft Launch and Recovery Equipment)
1421 Jefferson Davis Highway
Arlington, VA 22243-5120

Telephone: DSN 664-3355
Commercial, (703) 604-3355

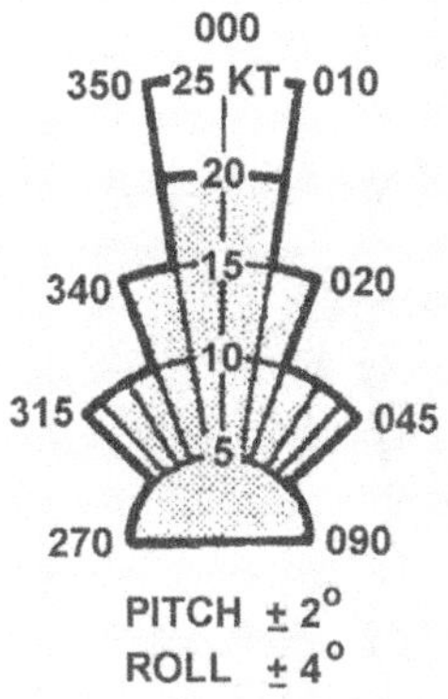

NOTES:

Helicopter aligned with ship's linup line and wind shown relative to the nose of the aircraft.

Entire envelope--day operations.
Shaded area--night operations.

Figure G-1. General launch and recovery wind limits

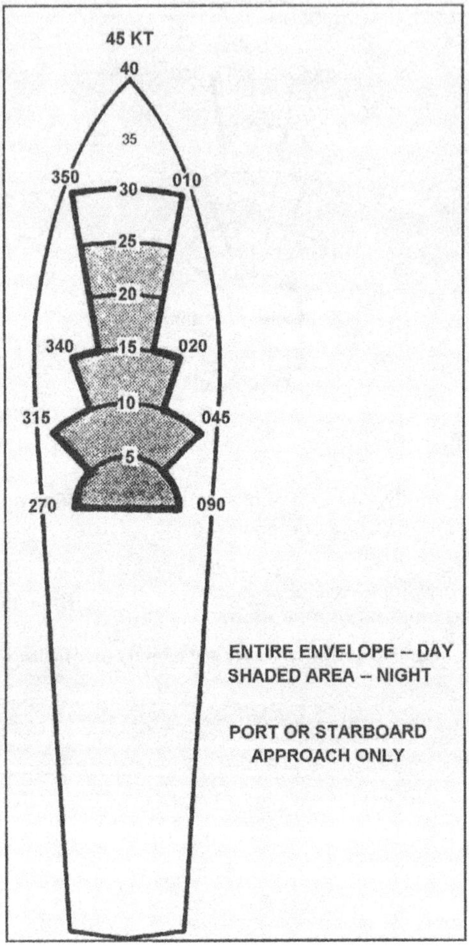

Figure G-2. General launch and recovery wind limits for LHA and LPH class ships

Section III. Typical Layouts, US Navy and USCG Ships

Figures G-3 through G-22 represent typical layouts of Navy and USCG ships as shown in NAEC-ENG-7576. These figures are for information purposly. The facilities resume must be consulted for the most current information on an assigned ship.

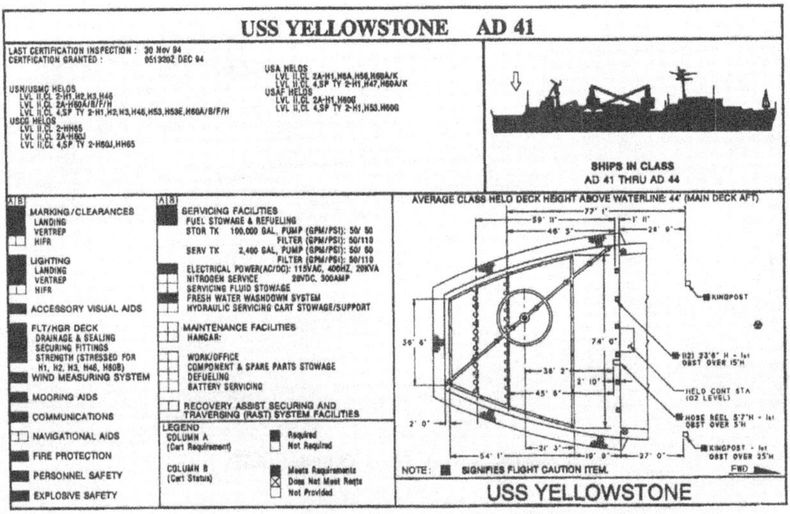

Figure G-3. Layout of the USS Yellowstone (AD 41)

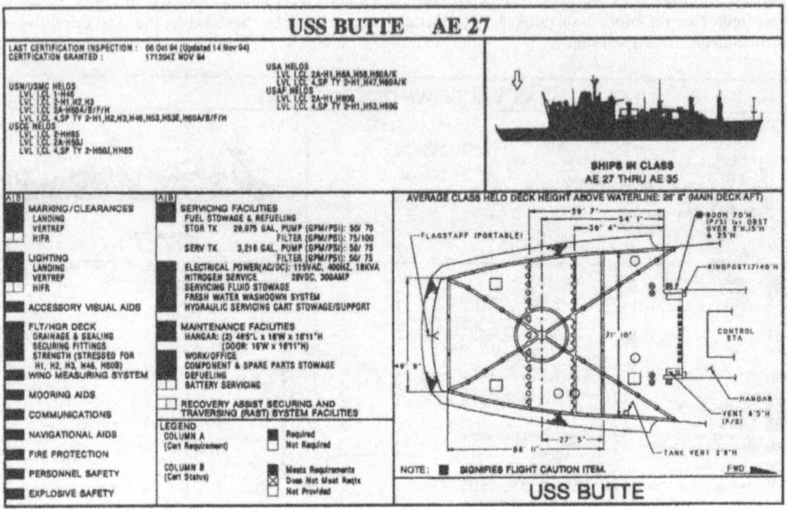

Figure G-4. Layout of the USS Butte (AE 27)

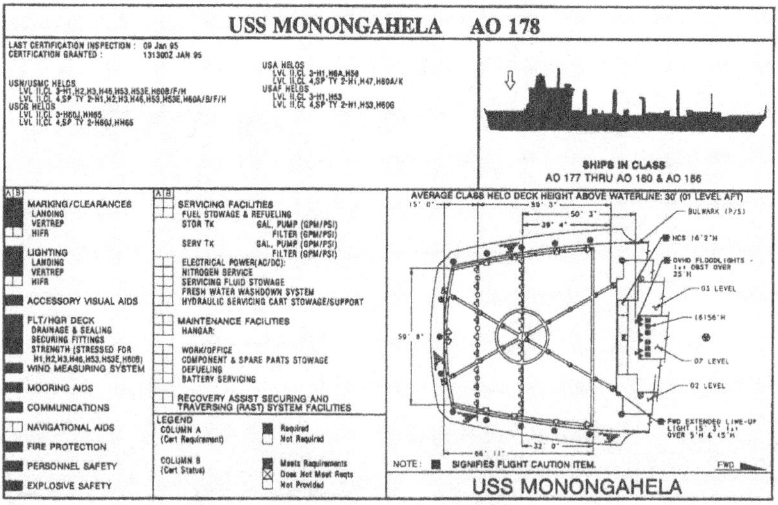

Figure G-5. Layout of the USS Monongahela (AO 178)

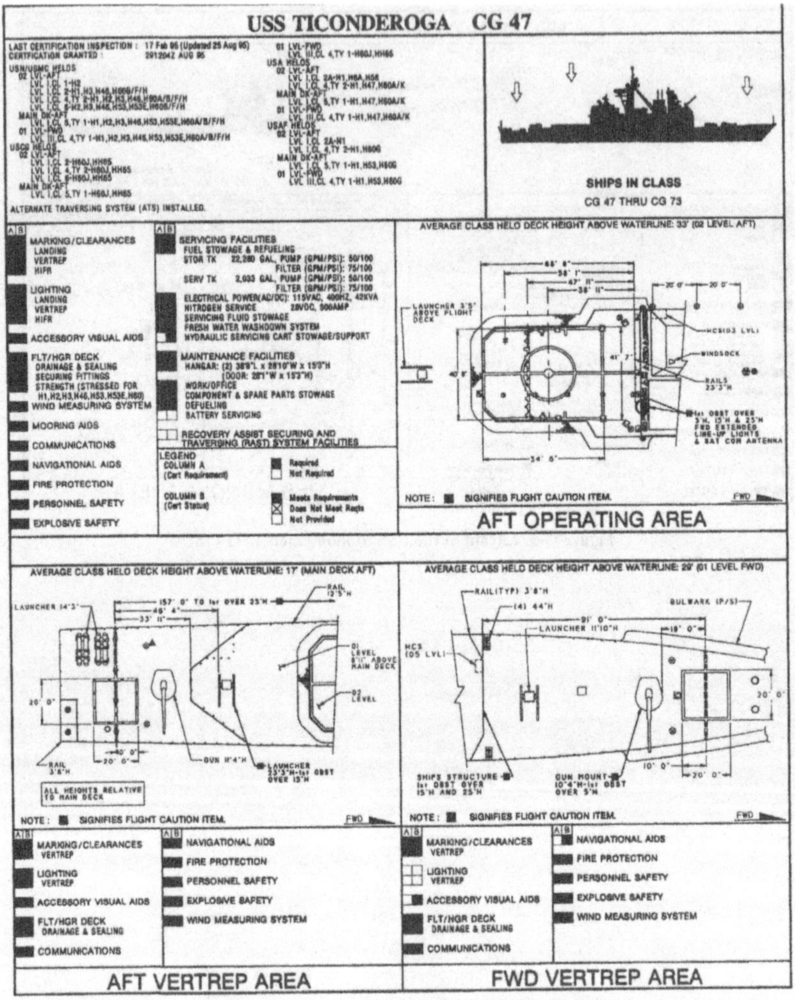

Figure G-6. Layout of the USS Ticonderoga (CG 47)

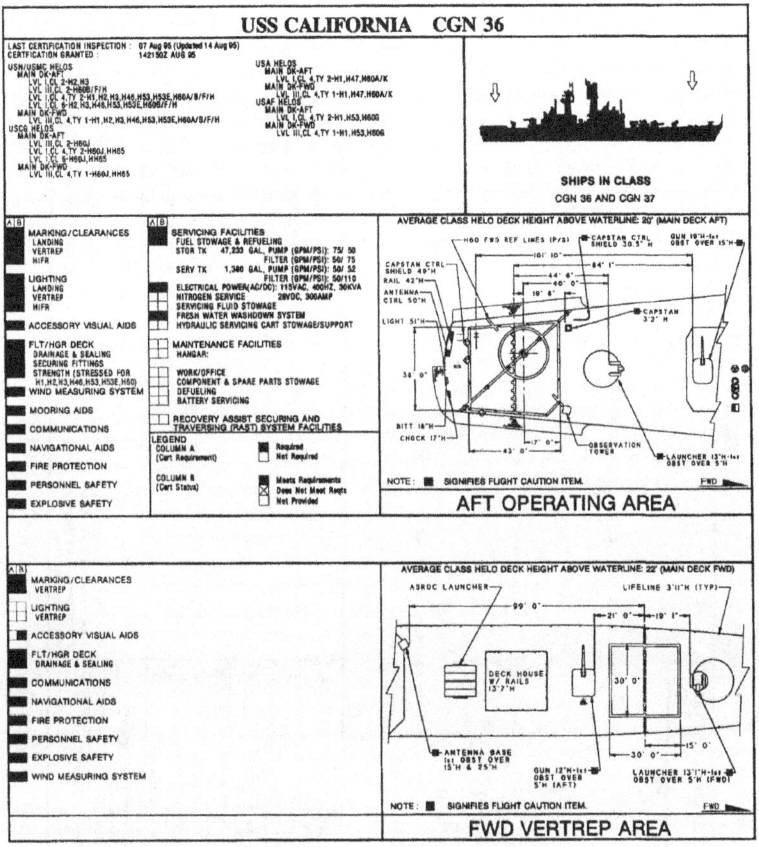

Figure G-7. Layout of the USS California (CGN 36)

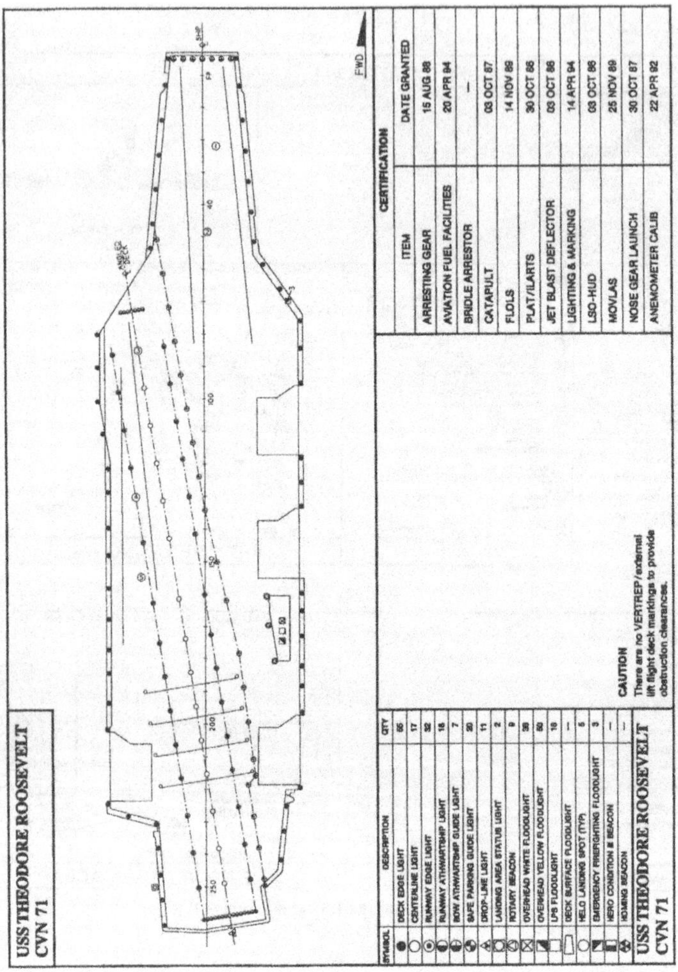

Figure G-8. Layout of the USS Theodore Roosevelt (CVN 71)

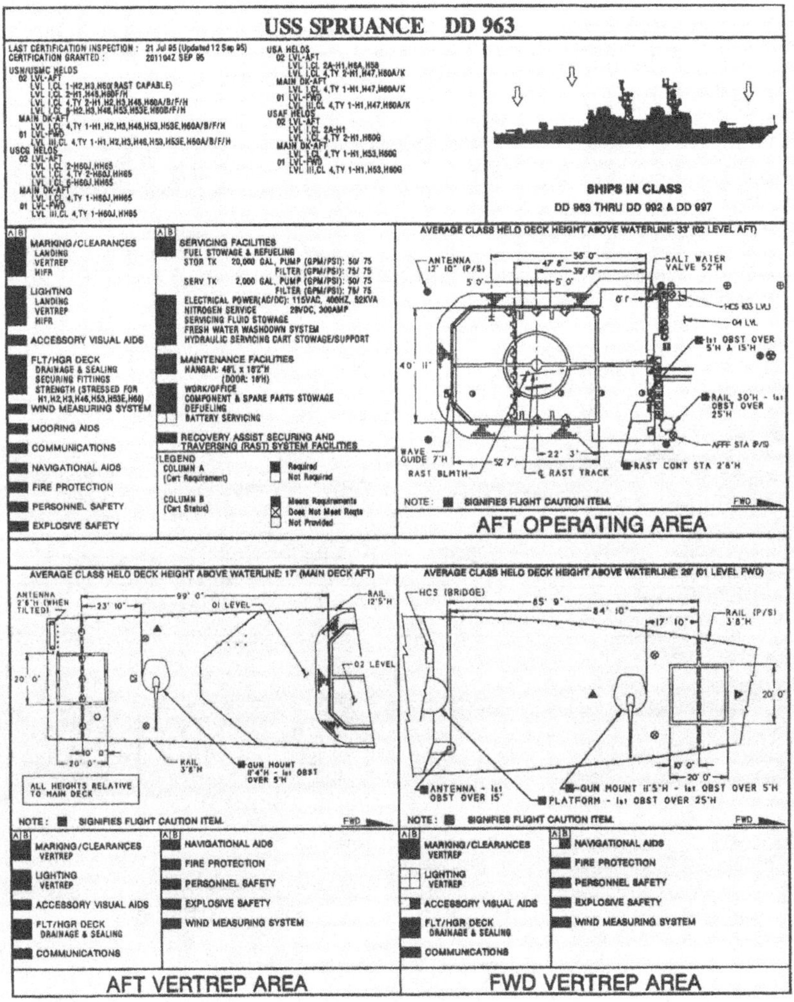

Figure G-9. Layout of the USS Spruance (DD 963)

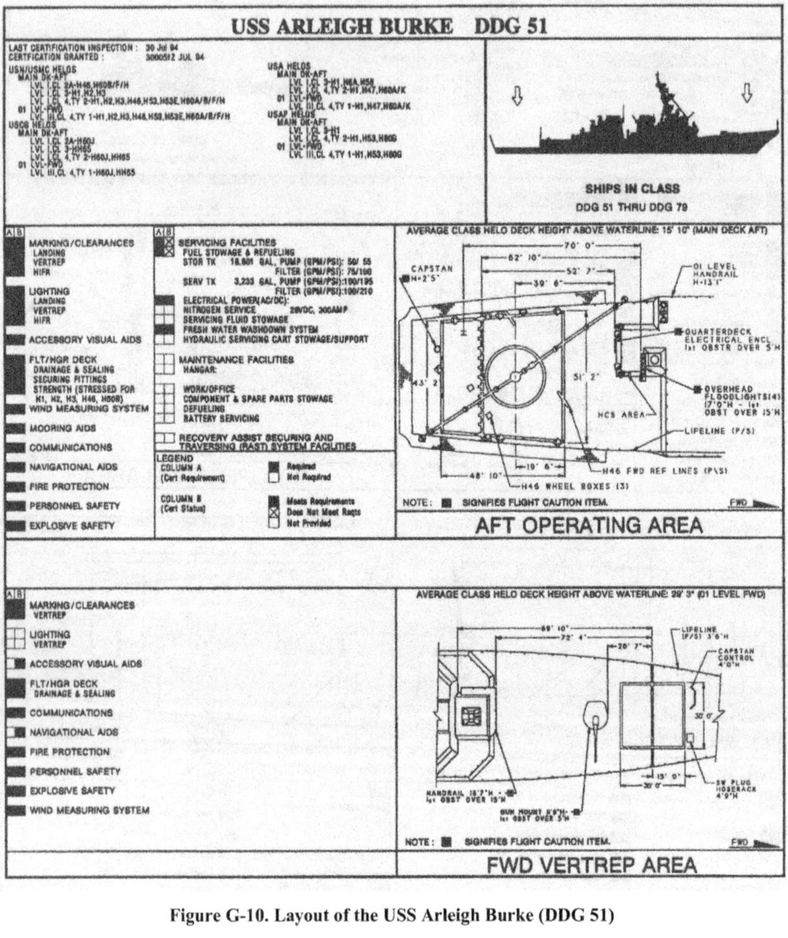

Figure G-10. Layout of the USS Arleigh Burke (DDG 51)

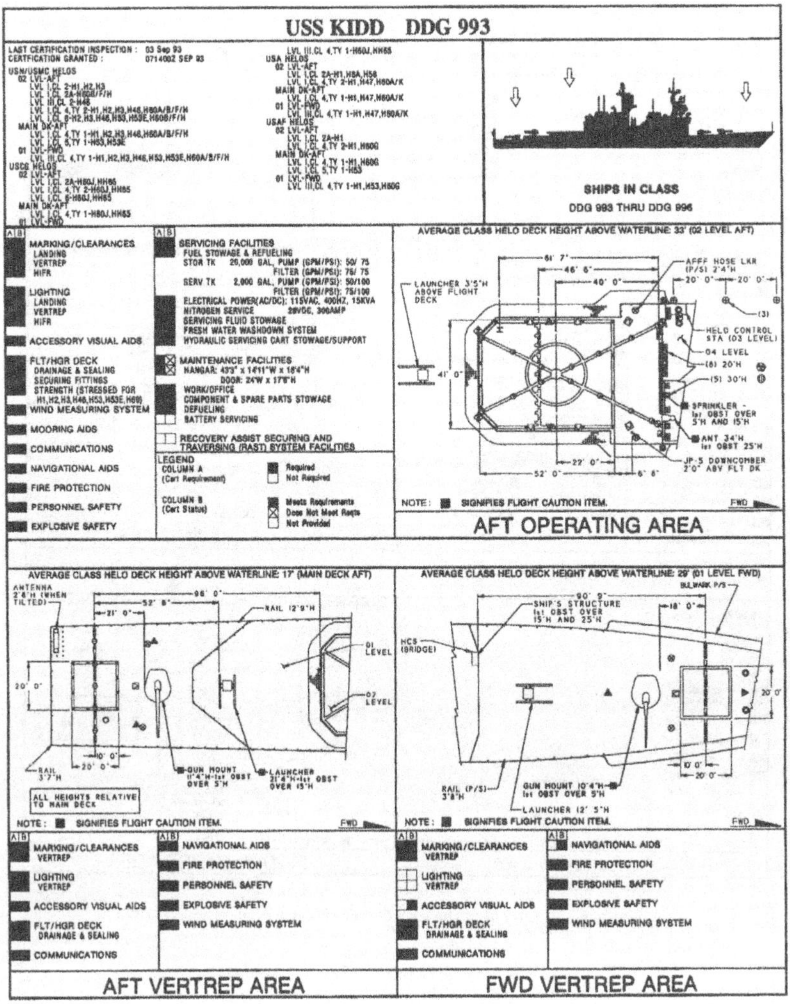

Figure G-11. Layout of the USS Kidd (DDG 993)

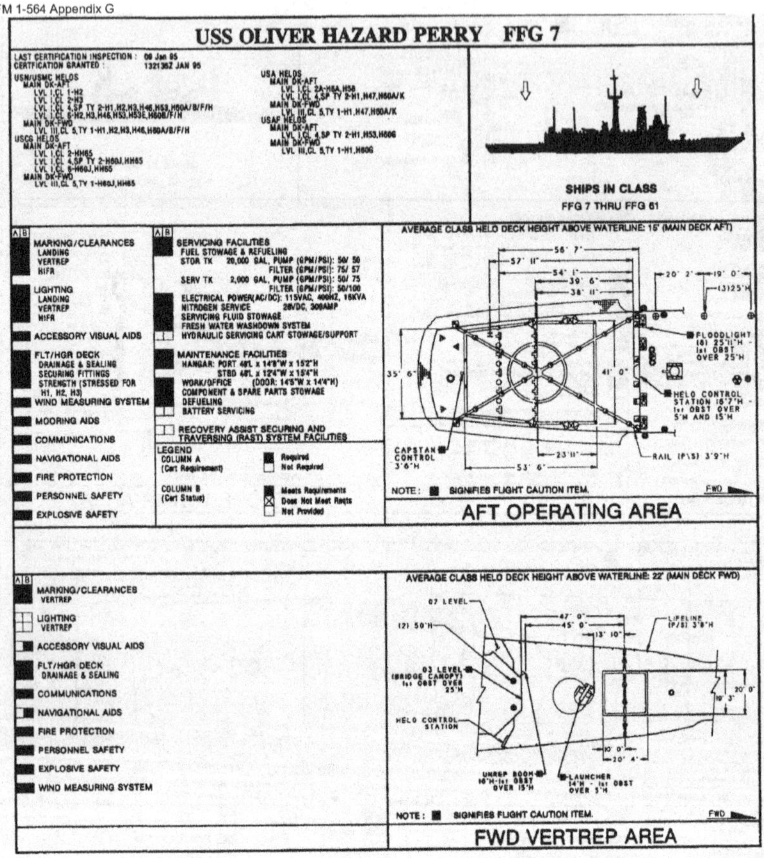

Figure G-12. Layout of the USS Oliver Hazard Perry (FFG 7)

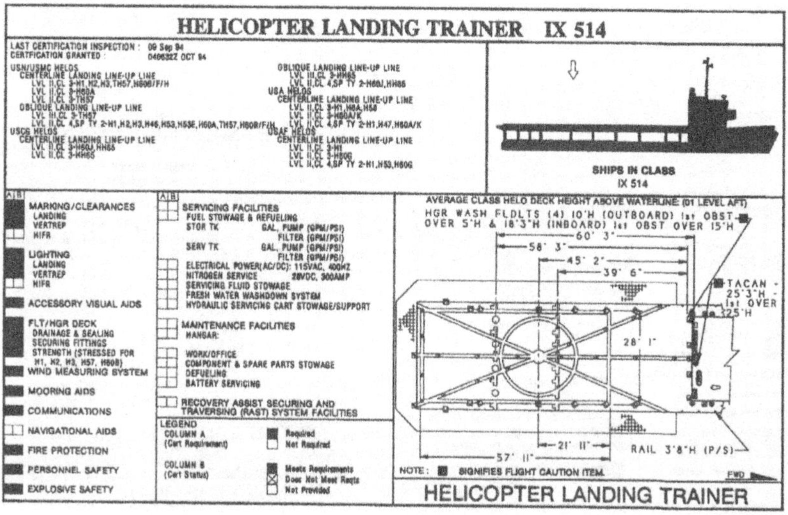

Figure G-13. Layout of the Helicopter Landing Trainer (IX 514)

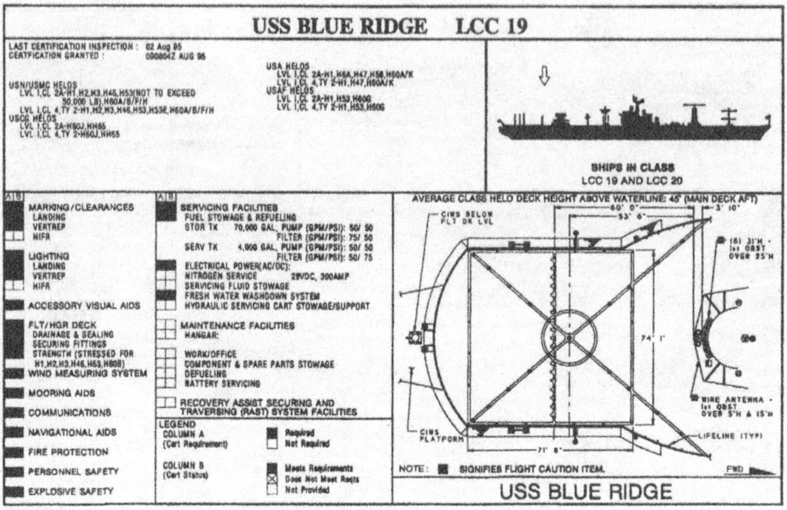

Figure G-14. Layout of the USS Blue Ridge (LCC 19)

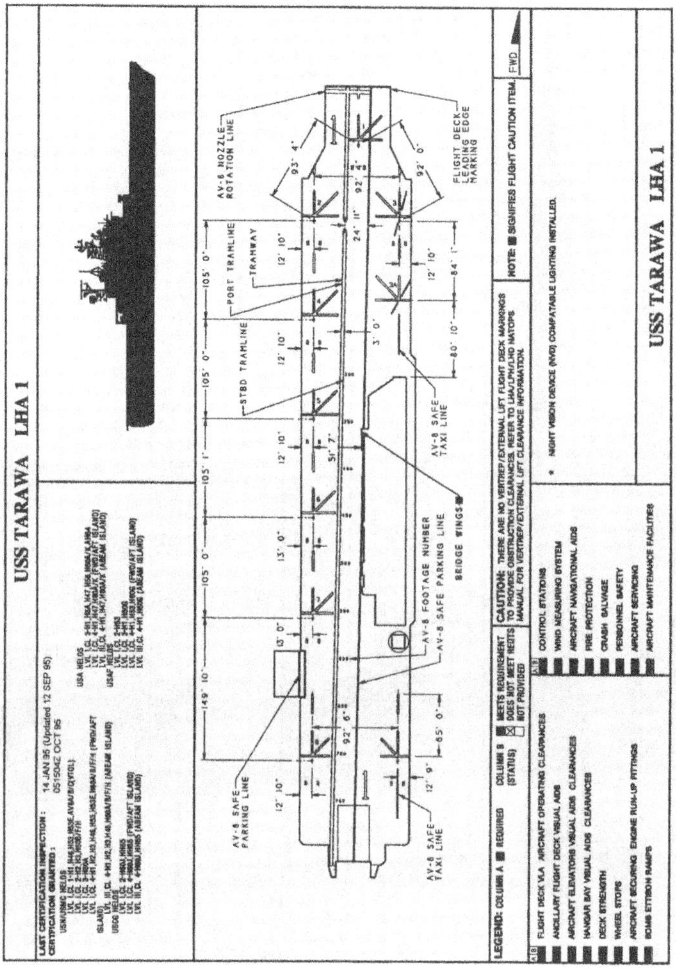

Figure G-15. Layout of the USS Tarawa (LHA 1)

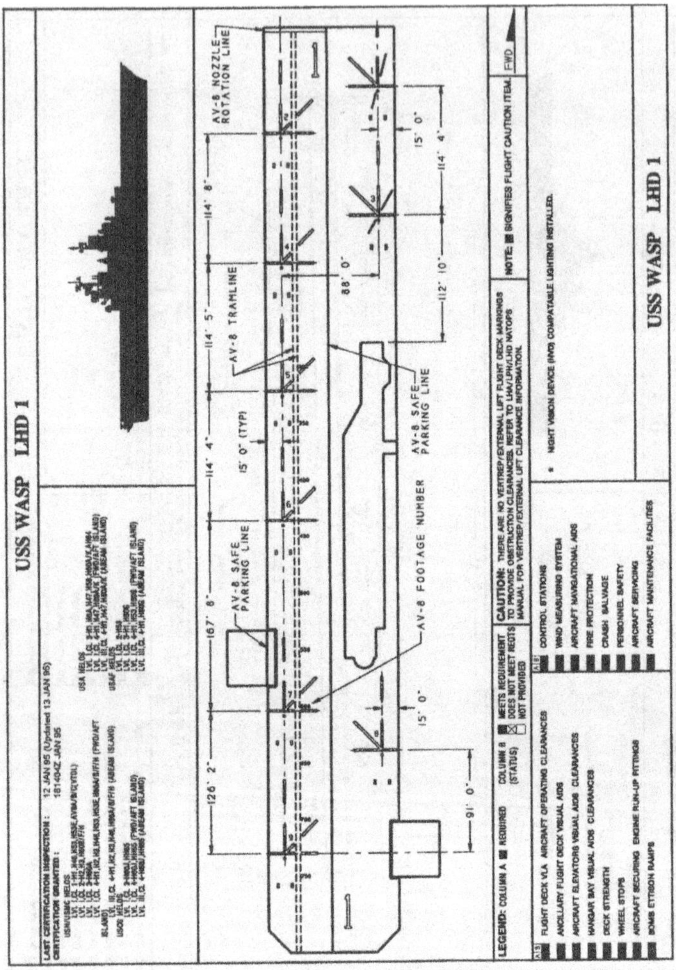

Figure G-15. Layout of the USS Wasp (LHD 1)

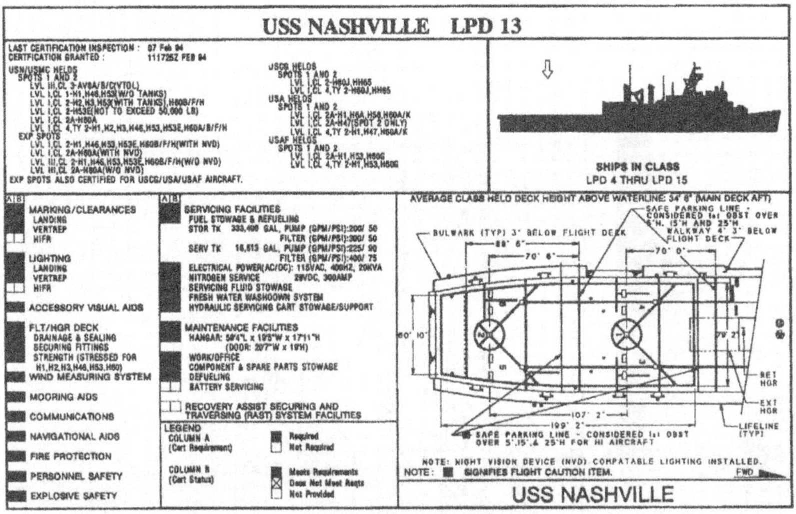

Figure G-17. Layout of the USS Nashville (LPD 13)

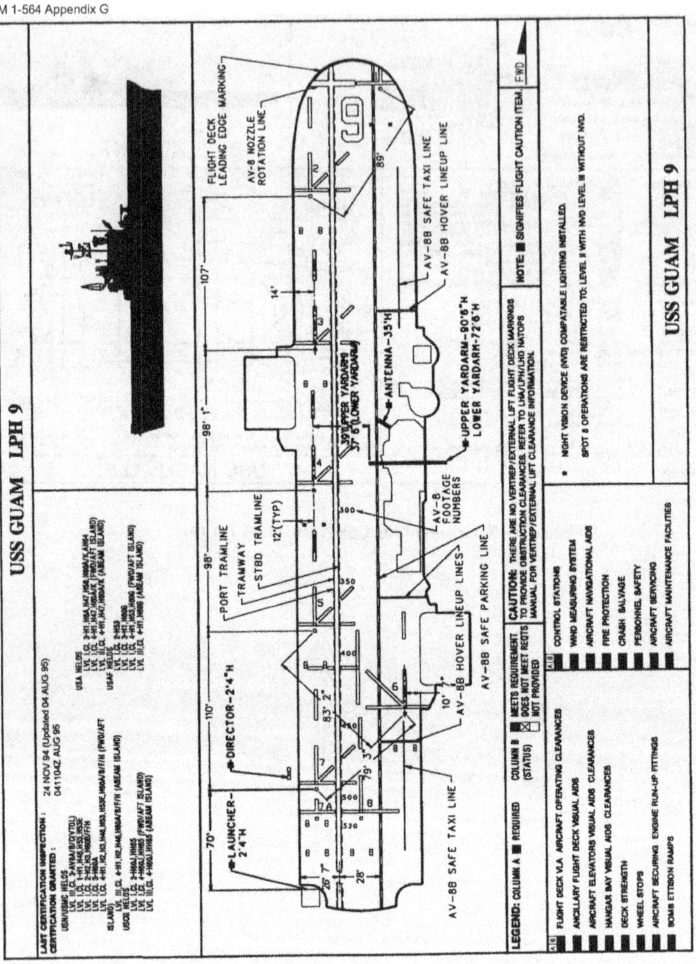

Figure G-18. Layout of the USS Guam (LPH 9)

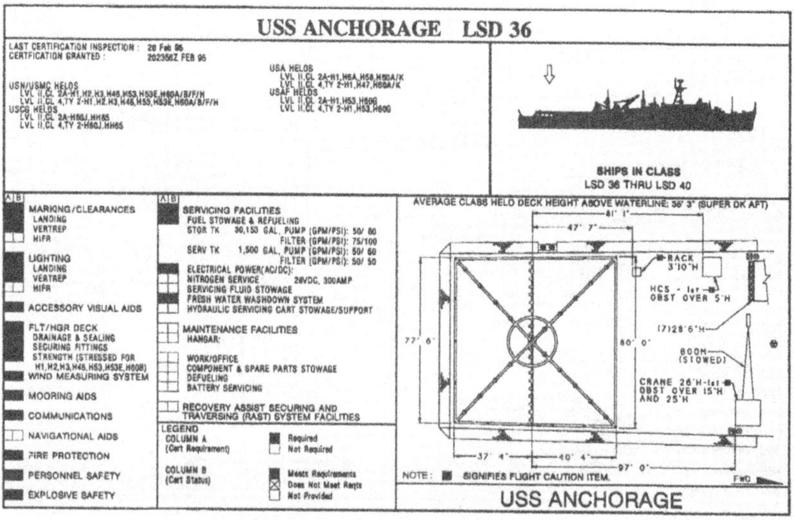

Figure G-19. Layout of the USS Anchorage (LSD 36)

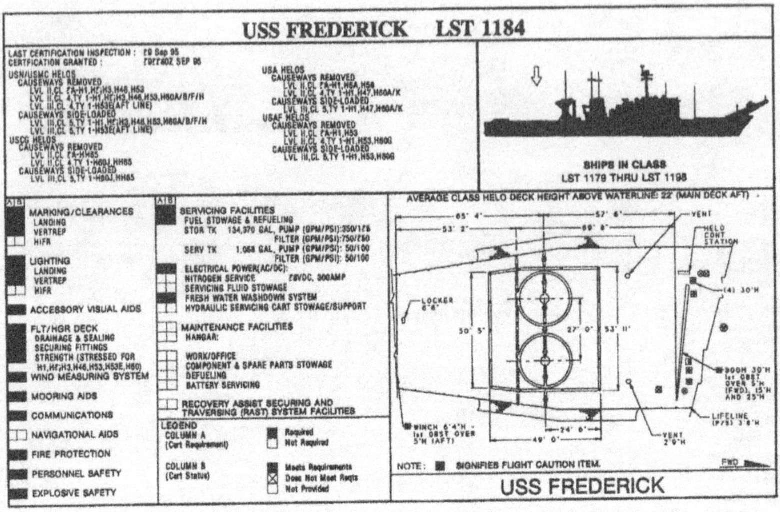

Figure G-20. Layout of the USS Frederick (LST 1184)

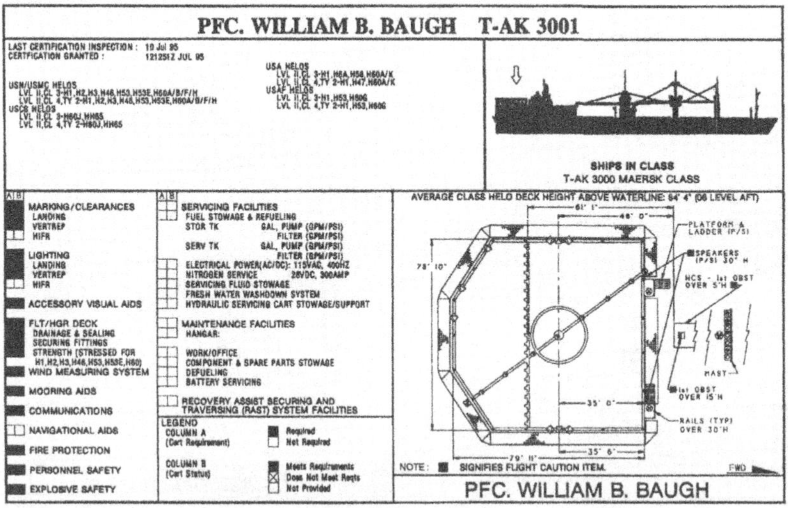

Figure G-21. Layout of the PFC. William B. Baugh (T-AK 3001)

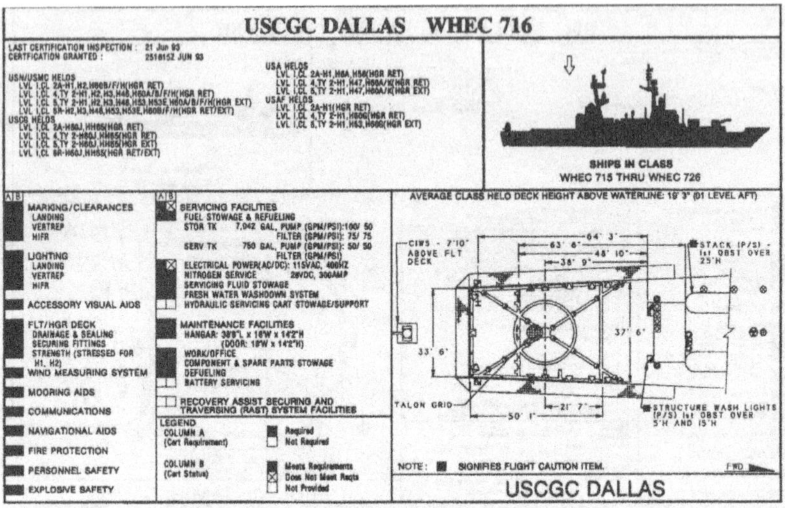

Figure G-22. Layout of the USCGC Dallas (WHEC 716)

APPENDIX H

MEMORANDUM OF UNDERSTANDING

The most current memorandum of understanding between the Army, Air Force, and Navy for deck landing operations was signed in 1988. The MOU (Figure H-1) is included in this manual for planning and information purposes. It will likely be updated during 1997-1998. Therefore, contact your MACOM to ensure that you have the most current MOU before planning shipboard operations.

MEMORANDUM OF UNDERSTANDING

Subject: Deck Landing Operations for US Army and US Air Force Helicopter Pilots

PURPOSE

1. To provide Department of the Navy, Department of the Army, and Department of the Air Force policy and procedures for Army and Air Force helicopter deck landing training and operations.

PURPOSE

2. The Departments of the Army and the Air Force receive deck landing training services in support of specific USA/USAF helicopter operations on US Navy ships. Support for a broader spectrum of joint operations and contingency plans may require joint training programs and Interservice Support Agreements.

DISCUSSION

3. Lessons learned from joint operations have shown the need for, and numerous contingency plans require, non-USN/USMC helicopters to operate from USN ships for Combat Search and Rescue (CSAR), combat support operations, medical evacuation (MEDEVAC), personnel transfer, and logistic support.

4. While there is no intent to provide deck landing training for all USA and USAF helicopter pilots who may be required to conduct operations on USN ships, this agreement approves developing a combined training program to qualify selected USA and USAF helicopter pilots for certain contingencies.

GENERAL PROVISIONS

5. ISAs and implementing procedures promulgated in a Letter of Instruction shall be provided by appropriate Naval, Army, and Air Force commands within the guidelines contained in the following attachments.

a. General - Attachment 1.

b. Initial qualification and currency requirements Attachment 2.

c. Aircrew Requirements - Attachment 3.

d. Ship certification and waiver - Attachment 4.

e. Forecasting and scheduling - Attachment 5.

f. Cost responsibility - Attachment 6.

g. References - Attachment 7.

h. Glossary - Attachment B.

i. Sample letter of instruction - Attachment 9.

6. Waivers and exceptions to the provisions outlined within this MOU will be handled on a case by case basis between USN (OP-593) and USA (DAMO-TRS) or USAF (XOOTA). Requests for waivers or exceptions should ve routed through appropriate service command channels for action.

7. This MOU shall remain in effect until amended by mutual written agreement between the Deparment of the Navy, the Department of the Army and the Department of the Air Force or until terminated in writing by the Department of the Navy, the Department of the Army, or the Department of the Air Force.

Figure H-1. Memorandum of understanding

GENERAL

1. All aviation operations shall be fully coordinated with each ship during a presail conference. Direct liaison is authorized (DIRLAUTH) between the ship and the participating US Army or US Air Force aviation unit.

2. Operations will be covered by an LOI published by the applicable US Navy type commander, and must be conducted IAW this MOU, the LOI, and references d thru j.

ATTACHMENT 1

INITIAL QUALIFICATION AND CURRENCY REQUIREMENTS

1. Single/dual (herein after referred to as single) spot ships:

   a. Initial day qualification consists of:

      (1) Flight training conducted by either a USA/USAF Deck Landing Qualification (DLQ) Instructor Pilot (IP) or a USN Helicopter Aircraft Commander (HAC) who is current on single-spot decks.

      (2) Ground School training per Attachment 3, para 5.

      (3) Six field deck landings prior to six single-spot shipboard landings, all within a ten consecutive day period.

   b. Currency requirements: Four single-spot shipboard landings within 90 days.

      (1) Pilots whose currency has lapsed, but have made four single-spot landings within the last 180 days, shall:

         (a) Undergo training conducted by either a current DLQ Pilot-in-Command (PC) or DLQ IP.

         (b) Perform four field deck landings prior to six shipboard landings, all within a ten consecutive day period.

      (2) Pilots whose currency has lapsed and have not made 4 single-spot landings within the last 181 days shall undergo initial qualification training.

   c. Night single-spot helicopter operations require significantly more training and specialized equipment than day operations and may not be conducted except for life-threatening emergencies or operational necessity. Requests for this type of training will not normally be approved. Exceptions will be handled on a case basis by USN (OP-593) and USA (DAMO-TRS), or USAF (XOOTA).

2. Multispot-spot ships (LPH/LHA/LHD):

   a. Initial day qualification consists of:

      (1) Flight training conducted by a USA/USAF IP or Unit Trainer (UT) who is day current or a current USN HAC.

      (2) Ground School training per Attachment 3, para 5.

      (3) Five day-field deck landings prior to five day-shipboard landings, all within a ten consecutive day period.

   b. Day currency requirements: Four shipboard landings within the preceding nine months. Pilots whose day currency has lapsed shall undergo initial day qualification; requalification shall be conducted by a USA IP, UT or PC. USAF requalification shall be conducted by a USAF IP.

c. Initial night qualification:

(1) The pilot shall be day-qualified and current.

(2) Ground School training per attachment 3, para 4.

(3) Flight training shall be conducted by a night-current USA/USAF DLQ IP or USN HAC.

(4) Six night-field deck landings prior to six night-shipboard landings, all within a ten consecutive day period. Pilots must also comply with the 72 hour requirement of para 2d.

d. Night currency requirements: Six night shipboard landings within the preceding 90 days are required to maintain currency. If more than 72 hours have elapsed since the last night shipboard landing, one day shipboard landing shall be performed within 24 hours prior to the next night shipboard landing.

3. Pilots qualified on single spot ships are qualified on multispot ships, but the reverse is not true.

4. Aircraft Carriers (CV): Routine DLQ training and operations normally will not be conducted on CV class ships. operations on CV class ships will be on a case basis and require a special ground brief by US Navy personnel, or Army/Air Force personnel designated by the Navy to give the briefing. Pilots qualified and current on single and multispot-spot ships shall be considered qualified and current on CV class ships.

5. Pilots performing Logistics-over-the-shore (LOTS) or vertical replenishment (VERTREP) operations that involve external loads without a shipboard landing shall be deck landing qualified and current. Pilots scheduled to participate in LOTS/VERTREP operations must receive a familiarization of the designated ship by US Navy Personnel or a previously familiarized US Army IP/PC or US Air Force IP/FE.

ATTACHMENT 2

AIRCREW REQUIREMENTS

1. U. S. Army:

a. Pilots shall be qualified and current IAW AR 95-1.

b. PC shall be deck landing qualified and current.

c. The pilot performing the deck landing shall be deck landing qualified and current, except when undergoing training.

2. U. S. Air Force:

a. Pilots shall be qualified and current IAW AFR 60-1.

b. The aircraft commander shall be deck landing current and qualified.

3. USN DLQ Instructors:

a. Shall be a DLQ current Helicopter Aircraft commander (HAC).

b. Shall be authorized by HQ DA (DAMO-TRS) or HQ USAF (XOOTA) to perform instructor pilot duties.

4. Ground School Training:

a. USA/USAF aircrews shall receive instruction to familiarize them with the mandatory operational procedures and training requirements for shipboard helicopter operations.

b. The Ground School course shall include but not be limited to:

(1) Landing Signals (Ref (k))

(2) Deck Markings (Refs (d) and (e))

(3) Emergency Procedures (Ref (e))

(4) Communications/NAVAIDs (Refs (d) and (e))

(5) Fuel/maintenance Support and Procedures (Ref (d))

(6) Landing Patterns/Approaches and Ship Control Zones (Refs (e), (f), and (g))

(7) VERTREP Procedures (if applicable) (Ref (e))

(8) Presail Conference Procedures/LOI

ATTACHMENT 3

Figure H-1. Memorandum of understanding (continued)

SHIP CERTIFICATION AND WAIVER

1. Waiver to conduct Army/Air Force helicopter operations is granted by Fleet Commanders, via Type Commanders (TYCOMs), on an individual basis.

2. Day VFR shipboard operations may be conducted by USA/USAF pilots on USN ships waivered for such operations.

3. Night VFR shipboard operations may be conducted by USA/USAF pilots on multi-spot ships waivered for such operations.

4. The Shipboard Aviation Facilities Resume (reference(d))lists all USN ships (including CVs); describes and depicts aircraft landing, VERTREP and hover facilities, flight deck marking, and lighting arrangements; and indicates US helicopters for which deck certification is granted.

ATTACHMENT 4

FORECASTING AND SCHEDULING

1. In order for the Navy to program for sufficient resources over the long term, estimates for annual training requirements will be provided to the Fleet commanders by Headquarters, Department of the Army (DAMO-TRS) and Department of the Air Force (XOOTA) with copy to the Chief of Naval Operations (OP-593F).

2. Normal training requirements shall be submitted in time for inclusion in the Fleet Commanders' quarterly employment scheduling conference.

3. Unscheduled training requirements shall be handled on a case by case basis in accordance with procedures established by the Fleet CINC's.

ATTACHMENT 5

Figure H-1. Memorandum of understanding (continued)

COST RESPONSIBILITY

1. The Navy shall schedule deck time to support USA/USAF DLQ training requirements. Demands for ship deck time beyond scheduled Fleet operations may require these ships services to be on a reimbursable basis.

2. The Army/Air Force shall provide the helicopters in which to conduct the training and will be responsible for helicopter operating costs.

3. Expenses for Navy personnel who are required to perform TAD/TDY in order to provide DLQ training for Army aviators/ Air Force pilots under the auspices of this MOU, will be borne by the US Army/US Air Force. When such travel is required, funding data shall be provided in advance of orders.

ATTACHMENT 6

REFERENCES

a. Unified Action Armed Forces (JCS Pub No. 2), October 1974.

b. Department of Defense Directive 4000.19, "Basic Policies and Principles for Interservices, Interdepartmental and Interagency Support Manual," October 1974.

c. Department of Defense Manual 4000.19 "Defense Retail Interservices Support Manual," October 1974.

d. NAVAIRENGCEN Pub 7576, "Shipboard Aviation Facilities Resume" - Lists all aviation capable ships, deck markings, fuel navigation aids and support facilities for US helicopters.

e. NWP-42, "Shipboard Helicopter Operating Procedures" Standardizes ground, flight, and operating procedures for single/double deck ships.

f. NAVAIR 00-80T-106, "LHA/LPH NATOPS Manual" -Standardizes ground, flight and operating procedures for LHA/LPH ships.

g. NAVAIR 00-8OT-105, "Cv NATOPS Manual" - Standardizes ground, flight and operating procedures for CV ships.

h. Army General Provisions and Flight Regulations(AR 95-1).

i. CNO Washington, DC 191736Z April 1983, Subj: Mixing Aviation Fuels.

j. Air Force Regulation 60-1, Flight Management, 28 May 1985.

k. NAVAIR 00-80T-113, "Aircraft Signals NATOPS Manuals" - Standard shipboard aircraft handling signals.

ATTACHMENT 7

GLOSSARY

Ship Certification - Certification of ships is categorized by "levels" and "classes" of required helicopter support facilities. The "level" identifies the environmental conditions under which the helicopter is authorized to operate and the "class" identifies the type of operations permitted and services available from the host ship.

Waiver - Authority granted by appropriate commanders to conduct operations and/or training not routinely authorized.

Aircrew - Pilots and crew members required to operate a helicopter.

Deck Landing Qualification (DLQ) Instructor Pilot (IP), Unit Trainer (UT), Pilot-in-Command (PC) - An IP, UT, or PC qualified and current in deck landings and designated in writing by the Unit Commander for such operations.

Shall - Indicates a mandatory procedure.

Operational Necessity - This term applies to missions associated with war or peacetime operations in which the consequences of an action justify the risk of loss of aircraft and aircrew.

Vertical Replenishment (VERTREP)/Logistics over The Shore (LOTS) - Missions involving the carrying of external loads/cargo to ships in which a landing is not anticipated.

Field Deck Landing - Practice shipboard landings performed at a shore facility with markings representative of a shipboard landing facility.

Deck Landing Qualifications - Landings performed on board ships for the purpose of qualifying pilots and aircrew for shipboard operations.

Pre-Sail Conference - Coordination meeting between host ship and operating unit for safety and operational planning.

Single/Dual-Spot Ships - Those certified ships having less than three adjacent landing areas.

Multi-Spot Ships - Those certified ships having three or more adjacent landing areas.

ATTACHMENT 8

SAMPLE LETTER OF INSTRUCTION

3120
Ser

From: Commander, Naval Surface Force, U. S.__________Fleet
To: Commanding Officer, USS Commander,

Subj: LETTER OF INSTRUCTION (LOI)

Ref: (a) [CINC EMPLOYMENT SCHEDULE]
(b) [SURFACE TYCOM DIRECTION TO CONDUCT US ARMY DLO'S]
(c) MOU between the DON and DOA/DOAF
(d) CNO Washington, DC 191736Z Apr 83 NOTAL

Encl: (1) Presail Conference Checklist

1. Summary. This LOI describes the concept of operations and assigns responsibility for Commanding officer, USS ________and assigned detachment from [US Army Command/US Air Force MACOM] for Deck Landing Qualification (DLQ) training exercises. This LOI is effective for planning for day/VFR operations. Night or IFR operations are not authorized.

2. Mission. USS ________will provide underway platform services in the conduct of DLQ training exercises (references (a) and (b) refer). Individual ship routine and exercises may be conducted consistent with attainment of DLQ training goals, safety, and operational security.

3. Concept of Operations. A detachment of helicopters from the US Army/US Air Force will conduct a series of DLQ training exercises consisting of day/VFR landings on the flight deck.

4. Command Relationships and Responsibilities:

a. COMNAVSURF ______is the officer-Scheduling-the-Exercise (OSE).

b. Commanding Officer, USS ________is assigned Officer-in-Tactical-Command (OTC) for scheduled DLQ training exercises and will coordinate with area/shore commands for appropriate OPAREA clearances. The host ship can provide limited administrative, logistics, material, maintenance and repair support. The OTC will ensure a flight deck safety/indoctrination brief is provided to US Army/US Air Force aircrews prior to the scheduled operations.

c. Officer-in-Charge of the [US Army helicopter detachment/US Air Force unit] is assigned as the Officer Conducting-the-Exercise (OCE) and is directed to conduct vigorous training exercises, pre-exercise training and planning, and to convene a presail briefing conference for major participants. The OCE will ensure prerequisites for shipboard helicopter operations are satisfied, will coordinate and supervise training exercises as they pertain to US Army/US Air

Force helicopters and will conduct appropriate preflight briefs.

5. Army/Air Force Helicopter Operations:

a. Shipboard Prerequisites for Helicopter Operations:

(1) Army/Air Force training requirements and personnel qualifications to conduct deck landing operations (reference (c)) will be attained prior to actual helicopter DLQ training exercises.

(2) US Army/US Air Force helicopter detachment will initiate coordination for a presail conference approximately four weeks prior to actual operation. A Presail Conference Checklist is provided at enclosure (1).

(3) U.S. Army/U.S. Air Force helicopter detachment will provide and officer on board ship for liaison between the ship and helicopter aircrews during DLQS.

(4) Army/Air Force liaison officer will provide diagrams of pertinent aircraft depicting aircraft egress, fuel cell locations, tie down points, and desired wind envelopes for the HCO and crash/fire crew during the presail conference.

(5) Army/Air Force aircrew personnel will brief flight deck/fire party personnel on Army/Air Force helicopter orientation/safety requirements to include a walk through of the aircraft. Aircraft orientation can be conducted following the first landing per type aircraft during DLQ operations.

(6) The low flash point temperature of JP-4/Jet B fuel constitutes a severe fire hazard aboard ship. In addition, JP-4/JP-5 mixtures assume the characteristics of JP-4 even with very low percentages of JP-4. To minimize the JP-4 hazard aboard ship, one of the following procedures shall be followed prior to first shipboard operation (listed in order of desirability):

(a) Defuel aircraft completely and refuel with JP-5, or

(b) Burn down to minimum fuel and refuel with JP-5 at least five (5) times.

JP-8/Jet A may be substituted for JP-5 in above procedures for shipboard flight operations. However, JP-8/jet A still reduces the flash point to alevel unacceptable for hangaring. Therefore, aircraft shall not be hangared if JP-8/Jet A has been substituted. Fuels other than JP-5 shall not be defueled into ship fuel systems. In an emergency, such fuels must be discharged directly over the side. Reference (d) provides additional discussion of fuel hazards.

(7) Waiver authority to conduct US Army/Us Air Force aircraft operations on US Navy ships must be granted by [FLEET CINC]. [SURFACE TYPE COMMANDER] will initiate the waiver request and inform all units concerned.

(8) Supported US Army/US Air Force units shall be familiar with pertinent US Navy shipboard aviation manuals:

(a) NAEC-ENG-7576 (Shipboard Aviation Facilities Resume); shows deck markings/ship profile.

(b) Pertinent sections of NWP-42 series (Shipboard Helicopter Operat-ing Procedures Manual): Launch/Recovery Procedures, Air Traffic Control, Aviation Fueling, General Helicopter Operations.

(9) Field deck landing requirements can be accomplished at Army/Air Force unit's home station if field deck markings are per Naval Air Engineering center specifications (available from [SURFACE TYPE COMMANDER]). Use of U. S. Naval field deck landing facilities at NAS Norfolk, VA, NAS Mayport, FL, NALF Imperial Beach, CA or NAS Barbers Point, HI, may be requested. Ample lead time (3-4 weeks) is required to ensure reservation for use.

6. Administrative/Logistics. The OCE (US Army detachment/US Air Force Unit Officer-in-Charge) is responsible for coordinating and arranging shore based administrative and logistics support. This support includes, but is not limited to, those suggested in item 22, enclosure (1).

7. Safety Reports. Actions to be taken in the event of aircraft mishap/incident will be per OPNAVINST 3750.6 (The Naval Aviation Safety Program). Initial message notification of aircraft mishap/incident will include as an information addressee, the [cognizant U. S. Army Headquarters: MESSAGE ADDRESS] or U. S. Air Force Headquarters: HQ USAF WASHINGTON DC //XOO//

ASSISTANT CHIEF OF STAFF
AVIATION

Copy to:
GROUP
SQUADRON
Participating Army/Air Force Unit(s)

ATTACHMENT 9

Figure H-1. Memorandum of understanding (continued)

SHIPBOARD HELICOPTER OPERATIONS
PRESAIL CONFERENCE CHECKLIST

	UNIT:	POC:	AV PHONE:
AVIATION:	________	________	________
SHIP:	________	________	________

1. ESTABLISH: a. DLQ Date:

____

b. Flight Schedule:

____

2. FIELD DECK LANDING QUAL RQMTS

(Ref: ARMY/AIR FORCE/NAVY MOU)

3. DLQ CURRENCY RQMTS

(Ref: ARMY/AIR FORCE/NAVY MOU)

4. TYPE AND NUMBER AIRCRAFT INVOLVED

5. # PILOTS NEEDING INITIAL QUAL/

CURRENCY (Ref: ARMY/AIR FORCE/
NAVY MOU)

6. SURFACE/AIR CLEARANCES

(Ship Responsibility)

7. AVIATION FACILITY WAIVER

(Type Commander (N8) Will
Coordinate)

8. TRANSIENT A/C LOCAL OPS

BRIEF (Base OPS Provides)

9. TACAN/RADIO FREQUENCIES

10. SHIP Overhead MSG (Containing

OPS/COMM Info)

11. SAFETY/OPERATIONS BRIEF(Ship/NWP 42)

12. CRASH RESCUE PROCEDURES AND POST

CRASH FIRE PROCEDURES

13. SEARCH AND RESCUE (SAR)

14. SHIPS GLIDE SLOPE INDICATOR (SGSI)

(DIFFERENT FROM ARMY GSI)

15. ENGAGE/DISENGAGE ENVELOPES

(Shipboard)

16. NO ROTOR BRAKES ON ARMY HELICOPTERS

17. TAKE OFF/RECOVERY ENVELOPES

18. BAD WEATHER PROCEDURES

19. FUEL REQUIREMENTS ON BOARD SHIPS

a. JP 5 Only

b. NATO 01 or Wiggins

20. FUEL REIMBURSEMENT (Standard

Military Credit Card or DD
Form 1348)

21. NAME OF ARMY/AIR FORCE

LIAISON OFFICER (During
Shipboard DLQ Period)

22. SHORE-BASED ADMINISTRATIVE/

LOGISTICS COORDINATOR

a. HELO RAMP PARKING*

b. FRESH WATER WASH

c. ACCOMMODATIONS

OFFICER

ENLISTED

d. MESS FACILITIES

e. LOCAL TRANSPORTATION

* Contact base Air Operations for transient parking and to obtain POC phone numbers for other logistic requirements.

NOTES:

COMNAVSURFLANT POC: N84 - COMM (804) 444-5340 or A/V 564-5340

COMNAVSURFPAC POC: N81 - COMM (619) 437-2393 or A/V 577-2393

Enclosure 1

Figure H-1. Memorandum of understanding (continued)

GLOSSARY

This glossary is divided into three sections. Section I contains definitions, Section II contains acronyms, and Section III contains brevity codes intended for air and ground operations personnel at the tactical level. The code words are for use during air-to-air, surface-to-air, and air-to-surface operations.

Section I. Definitions

airborne stores. Items to be carried internally or externally by aircraft (racks, launchers, adapters, and detachable pylons) that normally are not separated from the aircraft in flight (tanks, pods, guns, nonexpendable training weapons, and targets).

airborne weapons. Items to be carried internally or externally by aircraft that normally are separated from the aircraft in flight (missiles, rockets, bombs, mines, torpedoes, pyrotechnics, and ammunition).

air-capable ship. All ships other than CV/CVN or LPH/LHA/LHD from which aircraft can take off, be recovered, or routinely receive and transfer logistic support.

air operations. A section of the operations department that coordinates all matters pertaining to flight operations, including the proper function of AOCC/HDC.

air operations control center/helicopter direction center. A centralized air control agency that is responsible for the status keeping and tactical control of all aircraft not assigned to CIC/TACC. It also is responsible for IMC approach and departure control and becomes the helicopter direction center for tactical control of the helicopters during an amphibious operation.

air taxi. Jetborne or hovering flight at very low speed between two points.

amphibious assault aviation ship. An LPH, LHA, or LHD.

amphibious task force commander. The Navy officer designated in the initiating directive as commander of an amphibious task force.

angels. Altitude in thousands of feet.

approach control. A control station in AOCC/HDC that is responsible for controlling air traffic from marshal until hand-off to PriFly or the final controller. It also and also provides close control for all CCA wave-off traffic until a radar hand-off to another control station has been accomplished.

arming. An operation in which a weapon is changed from a safe condition to a state of readiness for initiation.

arming area. The area where ordnance is changed from a safe condition to a state of readiness. All arming evolutions required to be accomplished in the arming area by the aircraft stores loading manual/checklist shall be performed in this area. Before arming starts and before the aircraft take off, the area in front, behind, and/or surrounding the aircraft shall remain clear.

aviation ordnance evolution. A shipboard ordnance evolution requiring the breakout, buildup, and staging of ordnance and the loading, arming, launching, recovering, and dearming of ordnance-carrying aircraft.

aviation ship. A CV or CVN or an LHA, LHD, or LPH.

base recovery course. The ship's magnetic heading for aircraft recovery.

bingo. An order to an aircraft to proceed immediately to a divert field. Bearing, distance, and destination shall be provided.

bow. The front section of a ship or boat.

braking stop. The aft-most position of the nozzle control lever, that gives a component of reverse thrust on V/STOL aircraft.

buster. An order used by a ship controller to direct an aircraft to proceed at maximum speed.

carrier controlled approach. See precision approach.

center. A collective radio call for AOCC/HDC prefixed by a ship's code name that is used in the same manner as the shore-based counterpart.

charlie. A signal for aircraft to land aboard the ship. A number suffix indicates time delay in minutes before the landing may be anticipated.

cherubs. Friendly aircraft altitude in hundreds of feet.

clara. A pilot transmission meaning he does not have the visual landing aid (meatball) in sight.

close control. The tactical control of aircraft by a designated control unit whereby the aircraft receives orders affecting its movements. The pilot shall not deviate from instructions given him unless given clearance or unless unusual circumstances require him to take immediate action for the safety of the flight. In either case, the pilot shall inform the controller of the action taken. This type of control requires two-way radio communications and radar contact. The controller is responsible for the safety of the aircraft, and the pilot shall be informed whenever he is not held on the radarscope for more than 1 minute or five sweeps of the radar. The pilot is responsible for the ultimate safety of the aircraft.

control area. A circular airspace with a radius of 50 nautical miles around the ship that extends upward from the surface to unlimited an altitude and is under the cognizance of AOCC/HDC.

control zone. The airspace within a circular limit is defined by a 5 nautical miles horizontal radius from the ship. The control zone extends upward from the surface up to and including 2,500 feet unless otherwise designated for special operations and is under the cognizance of the air officer during VMC.

corrected hover weight. The thrust being used for takeoff or landing that is corrected for pressure, altitude, temperature, and individual engine characteristics of V/STOL aircraft.

cutback. Sudden and rapid reduction of engine speed as a result of JPT datum shift or dearming water switch during wet operations with V/STOL aircraft.

dearming (safing). An operation in which a weapon is changed from a state of readiness for initiation to a safe condition.

dearming area. The area where ordnance is changed from a state of readiness to a safe condition. All dearming evolutions are to be conducted in the dearming area by the individual stores loading manual/checklist. The area ahead, behind, and/or surrounding the aircraft shall be kept clear until all weapons and ordnance are completely safe. When taxiing aircraft from the landing area to the dearming area, care must be taken to reduce exposure of the armed ordnance to personnel and equipment.

deck status light. A three-colored light (red, amber, green) controlled from the primary flight control. For the Navy, the light displays the status of the ship to support flight operations. For the USCG, the light displays clearance for a helicopter to conduct a given evolution.

> Red deck status. The helicopter is not cleared for landing, takeoff, vertical replenishment, or helicopter in-flight refueling.
>
> Amber deck status. The helicopter is cleared to start engine(s) and engage or disengage rotors.
>
> Green deck status. The helicopter is cleared for landing, takeoff, vertical replenishment, or helicopter in-flight refueling.

delta. A signal given to hold and conserve fuel at an altitude and position appropriate to the type of aircraft and case recovery in effect. Also a pattern around the ship used to hold aircraft pending further clearance, assignment, and so forth.

density altitude. Pressure altitude in feet corrected for temperature and relative humidity. The higher the ambient air temperature and relative humidity, the higher the density altitude, which results in a decrease in aircraft performance.

departure control. A control station in AOCC/HDC that is responsible for the orderly flow of departing traffic.

downloading. An operation that removes airborne weapons/stores from an aircraft.

emergency expected approach time. A future time assigned before take off at which an aircraft is cleared to depart inbound or penetrate from a preassigned fix under lost communications conditions.

emergency final bearing. A magnetic heading that AOCC/HDC provides to all flight crews before take off to be used when executing emergency procedures for communications failure in IMC. The emergency marshaling pattern shall be relative to the EFB and is the final bearing for the lost communications TACAN approach.

emergency marshal. A marshal established by AOCC/HDC and assigned to each aircraft before take off. Emergency marshaling consists of a radial, DME, altitude, and emergency expected approach time.

emission control Control of all electromagnetic radiation, which includes electronic communications, radar, and visual systems. During its imposition, no electronic emitting device within the designated bands shall be operated unless absolutely essential to the mission of the force.

expected approach time. The future time at which an aircraft is cleared to depart inbound from a prearranged fix. Aircraft shall depart and start the approach at the assigned time if no further instructions are received.

father. TACAN.

feet dry. A pilot to AOCC/HDC report that indicates that the aircraft is passing over shore line and proceeding over land.

feet wet. A pilot to AOCC/HDC report that indicates the aircraft is passing the shore line and proceeding over water.

final bearing. The magnetic bearing assigned by AOCC/HDC for final approach. It is an extension of the landing area centerline.

final control. A control station in AOCC/HDC that controls traffic in instrument meteorological conditions until the pilot reports "VMC" or "meatball" or until he reaches approach minimums.

fleet area control and surveillance facility. A US Navy fixed, shore-based ATC facility. It is designated to manage offshore and inland operating areas and other assigned airspace, to include special-use airspace. The FACSFAC provides joint-use scheduling and control of surface, subsurface, and airborne military platforms operating within and transiting to and from these areas. It administers services to support the coexistence of military government and nongovenmrnent agencies consistent with national priorities.

flight deck officer. The officer responsible for the safe movement of aircraft on or about the flight deck of an aviation-capable ship.

flight level. Altitude expressed in hundreds of feet determined by setting 29.92 in the aircraft pressure altimeter; that is, 230 equals 23,000 feet in relation to the standard atmospheric pressure of 29.92.

flight quarters. A ship configuration that assigns and stations personnel at critical positions to conduct safe flight operations.

free-deck recovery. The launch or securing condition on the flight deck of a RAST-equipped ship when that system is not used.

ground resonance. A condition of geometric imbalance on helicopters caused by offset dynamic forces when the helicopter makes improper contact with the deck. If allowed to continue, destruction of the helicopter is imminent. Improper tie-downs aggravate the onset of ground resonance.

guard. A radio frequency that normally is used for emergency transmissions and is continuously monitored. (UHF: 243.0 MHz; VHF: 121.5 kHz.

helicopter control officer. In nonaviation facility ships, the helicopter control officer is responsible for supervising and directing launch and landing operations and for servicing and handling all embarked helicopters. Helicopter control officers will be graduates of the helicopter indoctrination course unless they are designated helicopter pilots.

helicopter direction center. See air operations control center/helicopter direction center.

HERO-safe ordnance. Any ordnance item that is sufficiently shielded or otherwise so protected that all EEDs/CADs contained by the item are immune to adverse effects (safety or reliability) when the item is employed in its expected shipboard RF environments provided the general HERO requirements are observed.

HERO-susceptible ordnance system. Any ordnance system proven (by tests) to contain EEDs/CADs that can be adversely affected by energy to the point that the safety and/or reliability of the system is in jeopardy when the system is employed in expected shipboard RF environments.

HERO-unsafe ordnance. Any ordnance item is defined as being HERO-unsafe when its external wiring is physically exposed; tests are being conducted on the item that result in additional electrical connections to the item; EEDs/CADs having exposed wire leads are present, handled, or loaded; the item is being assembled/disassembled; or the item is in a disassembled condition. Ordnance items that fall into the above classification may be exempted from being classified as HERO-unsafe ordnance as the result of HERO tests conducted to determine specific susceptibility.

hover. A condition of flight in which all movement relative to a fixed reference point has ceased.

hover stop. The position of the nozzle lever that vectors the thrust to the vertical position (81 degrees) on AV-8 aircraft.

hung weapons. Those weapons or stores on an aircraft that the pilot has attempted to drop or fire but could not because of a malfunction of the weapon, rack/launcher, or aircraft release and control system.

inbound bearing. The magnetic bearing assigned by AOCC/HDC that ensures interception of the final bearing at a specific distance from the ship.

India. IFF Mode IV.

instrument meteorological conditions. Meteorological conditions, when expressed in terms of visibility, distance from cloud and ceiling, are less than the minimal specified for visual meteorological conditions.

jetborne flight. Very slow speed flight supported by engine thrust only for V/STOL aircraft.

kilo report. A pilot report that indicates aircraft mission readiness. (Typically given to the controlling agency after takeoff.)

landing force commander. The officer designated in the initiating directive to command the landing force.

landing signal officer. The officer responsible for the visual control of aircraft in the terminal phase of the approach immediately before landing.

loading (rearming). An operation that installs airborne weapons and stores on or in an aircraft and may include fusing of bombs and stray voltage checks.

loading area. The area in which replenishment of airborne weapons or stores and other armament items on or in an aircraft is conducted. When handling weapons in this area, all fuse and initiators shall remain safe and all gun chambers shall be clear.

marshal. A bearing, distance, and altitude fix designated by AOCC/HDC from which pilots shall orient holding and from which initial approach shall begin.

marshal control. A control station in AOCC/ HDC that is responsible for the orderly flow of inbound traffic.

meatball (or ball). A pilot report that indicates that the visual landing aid is in sight, (amber beam of stabilized glide slope indicator).

medical evacuation. Evacuation of dead, wounded, sick, or otherwise incapacitated personnel by ship or air to an area or facility where the appropriate medical aid can be obtained.

mixed operations. Simultaneous V/STOL and helicopter air operations.

monitor control. The monitoring of radar and radio channels for emergency transmissions.

mother. Parent vessel (LHA, LPH, and LHD).

multispot ship. Ships that are certified to have three or more adjacent landing areas.

nonprecision approach. A radar-controlled approach or an approach flown by reference to navigation aids in which glide slope information is not available.

nonradar control. A form of air traffic control in which the pilot flies according to a published procedure or as prescribed by the controlling agency. Traffic separation is provided by the controlling agency using frequent pilot position reports and modified separation criteria. This form of control is used in case of an emergency, when all shipboard control radar is inoperative or, in the opinion of the AOCC/HDC officer, the shipboard control radar is unsafe.

operational necessity. A mission associated with war or peacetime operations in which the consequences of an action justify accepting the risk of loss of aircraft and crew.

parrot. Military IFF/transponder.

passengers/mail/cargo. An administrative/logistics flight scheduled for transfer of personnel and/or materiel to or from the ship. PMC does not include lifts of combat troops for actual or training vertical assaults or withdrawals.

pigeons. Magnetic bearing and distance from an aircraft to a specific location.

platform. A reporting point of 5,000 feet altitude in the approach pattern at which V/STOL aircraft reduce their rate of descent so as to arrive at 1,200 feet, 12 DME, and 250 knots.

pogo. A term used by a controlling agency indicating the return to the last assigned frequency if no contact is experienced on the newly assigned frequency.

popeye. A pilot term used to indicate that his aircraft has entered IMC.

port. The left hand side of a ship or boat (facing forward).

position and intended movement. The reference position of the officer in tactical command at a given time, and a forecast of the course and speed expected to be made during future movement. Position and intended movement are established to assist the return of the aircraft, to aid outlying surface units (pickets and so forth) in maintaining their stations, and for rendezvous purposes.

positive control. The tactical control of aircraft by a designated control unit, whereby the aircraft receives orders affecting its flight which immediately transfers responsibility for the safe navigation of the aircraft to the agency issuing such orders.

precision approach. An approach in which azimuth and glide slope information are provided to the pilot.

primary flight control. The controlling agency that is responsible for aircraft traffic control within the control zone.

ramp time. 1. The anticipated time designated by PriFly when the flight deck will be ready to recover aircraft. 2. The time the first aircraft in a Case III recovery is expected to be at the ramp.

rasberry. A ship-to-shore HF radio net used for flight following and administrative traffic concerning aircraft.

reaction controls. Variable exhaust ports at the extremities of the AV-8.

rearming area. The area where an operation is conducted that replenishes prescribed airborne weapons in or on an aircraft or where final dearming is accomplished following recovery and engine shutdown, or following ground abort. Only loading, down loading, arming and dearming evolutions are authorized to be conducted in the rearming area by the individual stores. All weapons handled or loaded in the rearming area shall be safe and remain safe.

safing (dearming). An operation whereby a weapon is changed from the state of readiness for initiation to a safe condition.

semi-jetborne flight. Flight where lift is provided by a combination of engine thrust and wing lift for V/STOL aircraft.

single-spot ship. Ships that are certified to have less than three adjacent landing areas.

spot. An approved shipboard helicopter landing site.

spin. A signal given to one or more V/STOL aircraft indicating a departure and reentry into the break. The command "spin" may be issued by either the air officer, LSO, or flight leader.

starboard. The right hand side of a ship or boat (facing forward).

stern. The rear section of a ship or boat.

strikedown. A term used to describe the movement of aircraft from the flight deck to the hangar deck level.

tactical air control center (afloat). The TACC is the primary air control agency within the amphibious objective area of responsibility from which all air operations supporting the amphibious force are controlled. This control refers to all airborne operations not incidental to the actual launch or recovery of aircraft: instrument departure, approach, and marshal.

tactical direction. A form of nonradar control in which tactical information is passed to an aircraft by the controlling unit. However, the aircraft commander is responsible for navigation and safety.

twelve nautical mile DME fix. A checkpoint in a CCA normally located on the final bearing, 12 miles from the ship. All V/STOL aircraft shall pass through the 12 nautical mile DME fix in level flight at an altitude of 1,200 feet and 250 KIAS and normally shall start transition to the landing configuration.

three nautical mile DME fix. A checkpoint in a CCA on the final bearing 3 miles from the ship through which all helicopters shall pass in a landing configuration.

transition. The maneuver of changing from nonconventional flight, wholly and partially jetborne, to conventional flight or, for V/STOL aircraft, vice versa.

trim back. The reduction of engine speed through action to hold constant at datum limit with V/STOL aircraft.

unexpended ordnance. Airborne ordnance that has not been subjected to attempts to fire or drop and is presumed to be in normal operating condition and can be fired or jettisoned if necessary.

vertical replenishment. The use of a helicopter for the transfer of material to or from a ship.

visual meteorological conditions. Weather conditions in which applies, expressed in terms of visibility, ceiling height, and aircraft clearance from clouds along the flight path. When these criteria do not exist, IMC prevails and IFR must be complied with.

V/STOL. An aircraft other than a helicopter whose characteristics of flight enable vertical and short takeoffs and landings.

warning. Operating procedures, practices, or conditions that may result in injury or death if not carefully observed or followed.

wave-off. An action to abort a landing initiated by PriFly, the LSO, the LSE, or the pilot at his discretion. The response to a wave-off signal is mandatory.

weather criteria requirements (helicopters).

Case I: 1,000-foot ceiling and 3 nautical miles visibility.

Case II: 500-foot ceiling and 1 nautical mile visibility.

Case III: Below 500-foot ceiling or less than 1 nautical mile visibility.

zip-lip. A condition that may be prescribed for flight operations during day or night VMC under which positive communications control is waived and radio transmissions are held to the minimum necessary for safety of flight.

Section II. Acronyms

AAA - antiaircraft artillery

ACHO - aircraft handling officer

ADIZ - air defense identification zone

AFFF - aqueous film forming foam

A/G - air-to-ground

AG - adjutant general

AGL - above ground level

AGM - air-to-ground missile

AIC - assign individual compressed dial

AL - Alabama

ALSE - aviation life support equipment

ammo - ammunition

AOA - angle of attack

AOCC - air operations control center

ARSOA - Army special operations aviation

ARTCC - air route traffic control center

ASM - antiship missile

ASR - approach surveillance radar

ASW - antisubmarine warfare

ATAS - air-to-air Stinger

ATC - air traffic control

ATO - air transportation officer

ATM - assign traffic metering

attn - attention

AWACS - Airborne Warning and Control System

AWR - air worthiness release

BRC - base recovery course

C^2 - command and control

CA - California

CAI - close-in approach indicator

cal - caliber

CAP - combat air patrol

CATF - commander, amphibious task force

CCA - carrier controlled approach

CCO - combat cargo officer

CECOM - (Army) Communications-Electronics Command

CH - cargo helicopter

CIC - combat information center

CO - commanding officer

comm - communications

COMSEC - communications security

CPO - chief petty officer

CQ - carrier qualification

CTF - commander, task force

CV - aircraft carrier

CVN - nuclear powered attack aircraft carrier

DA - Department of the Army

DC - District of Columbia

DD - Department of Defense

DF - direction finding

DIRLAUTH - direct language authorized

DLQ - deck landing qualification

DME - distance measuring equipment

DSN - Defense Switching Network

EAT - expected approach time

EEAT - emergency expected approach time

EED - electroexplosive device

EEFI - essential elements of friendly information

EFB - emergency final bearing

EMCON - electronic emission control

EO - elevator operator

EOD - explosive ordnance disposal

ESSS - external support stores system

ETE - estimated time en route

FAA - Federal Aviation Administration

FACSFAC - fleet area control and surveillance facility

FAC - floor of controlled airspace

FAC-A - forward air controller--airborne

FAF - final approach fix

FCLP - field carrier landing practice

FDLP - field deck landing practice

FDO - flight deck officer

FFG - guided missile frigate

FL - Florida; flight level

FLIP - flight information publication

FM - frequency modulated; field manual

FOD - foreign object damage

FORSCOM - (United States Army) Forces Command

FRAGO - fragmentary order

ft - foot; feet

GAIL - glide angle indicator light

GCA - ground controlled approach

GCI - ground control interception

GSI - glide slope indicator

GPS - global positioning system

GTS - gas turbine starter (V/STOL)

HAC - helicopter aircraft commander

HAPI - horizontal approach path indicator

HARM - high altitude antiradiation missile

HCO - helicopter control officer

HDC - helicopter direction center

HEEDS - helicopter emergency egress device system

HEI - high-explosive incendiary

HERO - hazards of electromagnetic radiation to ordnance

HF - high frequency

HLT - helicopter landing trainer

HPI - hover position indicator

HPRF - high pulse radar frequency

HQ - headquarters

HUD - heads-up display

HVAA - high value airborne assets

I - individual

IAF - initial approach fix

ID - identification

IFF - identification, friend or foe

IMA - intermediate maintenance activity

IMC - instrument meteorological conditions

IN - instructor

info - information

INS - inertial navigation system

IP - instructor pilot

IR - infrared

JFC - joint forces commander

JPT - jet pipe temperature (V/STOL)

JPTL - jet pipe temperature limiter (V/STOL)

JTF - joint task force

KIAS - knots indicated airspeed

kt - knot

LDNS - lightweight doppler navigation system

LHA - amphibious assault ship

LHD - amphibious assault ship

LOI - letter of instruction

LOS - line of sight

LOTS - logistics over the shore

LOX - liquid oxygen

LPH - amphibious assault ship

LPO - lead petty officer

LPU - life preserver unit

LSE - landing signalman enlisted

LSO - landing signal officer

MACOM - major Army command

mar - marshal

MDA - minimum decent altitude

MEDEVAC - medical evacuation

MIM - maintenance instruction manual

min - minute

mm - millimeter

MOA - memorandum of agreement

MOU - memorandum of understanding

MPRF - medium pulse radar frequency

msg - message

MSL - mean sea level

NATO - North Atlantic Treaty Organization

NAS - naval air station

NAVAID - navigational aid

NAVSEA - naval sea

NC - North Carolina

NCOIC - noncommissioned officer in charge

NCTR - noncooperative target recognition

NDB - nondirectional beacon

NFO - naval flight officer

NJ - New Jersey

nm - nautical mile

NSN - national stock number

NVD - night vision device

NWP - naval warfare publication

NWS - nosewheel steering

OBA - oxygen breathing apparatus

OCE - officer conducting exercise

ODO - operations duty officer

OIC - officer in charge

OOD - officer of the deck

ops - operations

OTC - officer in tactical command

PA - Pennsylvania

PAR - precision approach radar

PAX - passengers

PC - pilot in command

PIM - position and intended movement

PMC - passengers, mail, or cargo

POC - point of contact

PP - power plant

PriFly - Primary Flight (Control)

QA - quality assurance

qual - qualification

RADHAZ - radiation hazard

RAST - recovery assistance, securing, and training system

rel - relative

ROE - rules of engagement

RPM limit - fan speed limit (V/STOL)

RPV - remotely piloted vehicle

rqmt - requirement

RVL - rolling vertical landing (V/STOL)

RVTO - rolling vertical takeoff (V/STOL)

RWR - radar warning receiver

SAAHS - stability augmentation attitude hold system

SAM - surface-to-air missile

SAR - search and rescue

SAS - stability augmentation system

SE - southeast

SEAD - suppression of enemy air defense

SF - standard form

SGSI - stabilized glide slope indicator

SIF - selective identification feature

SIGINT - signal intelligence

SGSI - stabilized glide slope indicator

SO - safety observer

SOP - standing operating procedure

SRC - special requirements code

STO - short takeoff (V/STOL)

tac - tactical

TACAN - tactical airborne navigation

TACC - tactical air control center

TACRONS - tactical air control squadron

TAO - tactical action officer

TBM - tactical ballistic missile

TCA - tactical control area

TDY - temporary duty

TL - team leader

TM - team member

TRADOC - (United States Army) Training and Doctrine Command

TYCOM - type commander

UAV - unmanned aerial vehicle

UIC - unit identification code

UH - utility helicopter

US - United States

USA - United States Army

USAF - United States Air Force

USMC - United States Marine Corps

USN - United States Navy

UT - unit trainer

VA - Virginia

VERTREP - vertical replenishment

VFR - visual flight rules

VLA - visual landing aid

VMC - visual meteorological conditions

V/STOL - vertical and/or short takeoff and landing

WOD - wind over deck

XO - executive officer

yd - yard

Section III. Multiservice Brevity Codes

abort A directive to cease the action, attack, event, or mission.

action A directive to initiate a briefed attack sequence or maneuver.

active An emitter is radiating.

alarm A directive or information indicating the termination of EMCON procedures.

alligator Link-11/TADIL A.

alpha check A request for bearing and range to a described point.

anchor(ed) 1. Orbit about a specific point; refueling track flown by tanker.

2. Information to indicate a turning engagement about a specific location.

angels Height of friendly aircraft in thousands of feet.

Arizona No arm ordnance remaining.

as fragged Unit or element will be performing exactly as stated by the air tasking order.

authenticate To request or provide a response for a coded challenge.

autocat Any communications relay using automatic retransmissions.

autonomous Aircrew is operating without the benefit of GCI, AIC, or AWACS.

azimuth Two groups separated in bearing.

bandit A positively identified enemy aircraft. The term is a function of identification and does not necessarily imply direction or authority to engage.

banzai Information or directive to execute launch and defend tactics.

base (number) Reference number used to indicate such information as headings altitude, fuels, and so on.

bead window Last transmission potentially disclosed unauthorized information.

beam(ing) Target maneuvering stabilized within 70- to 110-degree aspect; generally given with cardinal directions such as east, west, north, and south.

belly check A momentary unloaded bank to check the blind side of a turning aircraft.

bingo Prebriefed fuel state which is needed for recovery using prebriefed parameters.

bird Friendly surface-to-air missile.

bird(s) affirm Fire control radar is locked on designated target.

bird(s) away Missile has been fired at designated target.

bittersweet Notification of possible blue-on-blue situation relative to a designated track or friendly aircraft.

blank A SEAD aircraft does not detect any emitters of interest.

blind No visual contact with friendly aircraft or ground position; opposite of the term "visual."

blow through A directive or informative call that indicates aircraft will continue straight ahead at the merge and not turn with target.

bogey Unidentified air contact.

bogey dope Request for target information as briefed or available.

box Groups, contacts, or formations in a square or offset square.

braa Format of tactical control providing bearing, range, altitude, and aspect from fighter.

bracket Indicates geometry where friendly aircraft will maneuver to a position on opposing sides, either laterally or vertically from the target.

break (direction) Directive to perform an immediate maximum performance turn in the indicated direction. Assumes a defensive situation.

breakaway Tanker or receiver call indicating immediate vertical and nose or tail separation between tanker and receiver is required.

brevity Term used to denote radio frequency is becoming saturated or degraded and briefer transmissions must follow.

broadcast Request or directive to switch to broadcast control.

broke lock Loss of radar or IR lock-on (advisory).

buddy lock Locked to a known friendly aircraft. Normally a response to a "spiked" or "buddy spiked" call and accompanied with "angles or altitude."

buddy spike Friendly aircraft air-to-air indication on RWR. To be followed by position, heading, and altitude.

bugout (direction) Separation from that particular engagement/attack/operation; no intent to engage or return.

bulldog Friendly antiship cruise missile.

bullseye An established point from which the position of an aircraft can be referenced. Made by cardinal/range or digital format.

bump/bump up A fly-up to acquire LOS to the target or laser designation.

buster A directive call to fly at maximum continuous speed (military power).

buzzer Electronic communications jamming.

burner A directive to select or deselect afterburner.

cap/capping 1. A directive call to establish an orbit at a specified location.

2. An orbit at an specified location.

captured The aircrew has identified and is able to track a specified A/G target with an onboard sensor.

cease engagement Break the engagement on the target specified and prepare to engage another target. Missiles in flight will continue to intercept.

cease fire Do not open fire or discontinue firing; complete intercept if weapons are in flight; continue to track.

champagne An attack of three distinct groups with two in front and one behind.

chattermark Begin using briefed radio procedures to counter communications jamming.

cheap shot 1. (USAF) Active missile not supported to active range or medium PRF.

2. (USN) Active missile not supported to active range.

check Turn ( ) degrees left or right and maintain new heading.
(left or right)

cherubs Friendly aircraft altitude in hundreds of feet.

chicks Friendly aircraft.

clean 1. No radar contacts on bandits, bogies, or aircraft of interest.

2. No visible battle damage.

cleared Requested action is authorized (no engaged/support roles are assumed).

cleared hot Ordnance release is authorized.

cloak Directive or informative switch from normal external lighting to covert NVD-only compatible lighting.

closing Decreasing in range.

cold 1. Attack geometry will result in a pass or roll-out behind the target.

2. On a leg of the CAP pointed away from the anticipated threats.

3. Threat group heading away from fighters. Opposite of "hot."

come off A directive to maneuver as indicated to either regain mutual support or to

(left/right/low/dry) deconflict flight paths for an exchange of engaged and supporting roles. Implies both "visual" and "tally."

committed/ commit Fighter intent to engage/intercept; controller continues to provide information.

confetti Chaff lane or corridor.

cons/conning Threat aircraft leaving contrails.

contact 1. Sensor contact at the stated position.

2. Acknowledges the sighting of a specified reference point.

continue Continue present maneuver; does not imply clearance to engage or expend ordnance.

cover 1. Directive to initiate S/A engagement on specified track up to the point of firing.

2. Assume a posture that will allow engagement of the specified target if directed.

crank (direction) F-pole maneuver; implies illuminating target at radar gimbal limits.

cutoff Request for or directive to intercept using cutoff geometry.

cyclops Any UAV/RPV.

dash (number) Aircraft position within a flight. Same as chalk. Use if a specific call sign is unknown.

deadeye Informative call by an airborne laser designator indicating the laser/IR system is inoperative.

declare Inquiry by fighter to an AWACS, a GCI, an AIC, or a capable aircraft as to the identification of a correlated group.

defensive (Spike, missile, SAM, Mud, or AAA) Aircraft is in a defensive position and maneuvering with reference to the stated condition.

deploy Directive for the flight to maneuver to briefed positioning.

divert Proceed to alternate mission or base.

dolly Link-4A/TADIL C.

drag/dragging 1. (USAF) Target maneuvering 0 to 60-degree aspect.

(direction) 2. (USN) Target maneuvering to 120- to 18- degree aspect.

drop/dropping Directive or information to stop monitoring a specified emitter or target and resume search responsibilities.

duck Tactical air-launched decoy.

echelon (direction) Groups, contacts, or formation with wingman displaced approximately 45 degrees behind leader's 3/9 line.

echo Positive seesaw, EWWS, system M/Mode X reply.

engaged Maneuvering with the intent of achieving a kill. If no additional information is provided (bearing or range), this implies visual/radar acquisition of target.

estimate Using information available to provide data required; implies degradation.

extend (direction) Short term maneuver to gain energy and distance or separation normally with the intent of reengaging.

eyeball 1. Fighter with primary visual identification responsibility.

2. EO/IR acquisition of an aircraft.

fade Directive call to HVAA to continue present mission while extending range from target in response to perceived threat.

faded Previous radar contact is lost.

fast Target speed is estimated to be 600 knots ground speed/Mach 1 or greater.

father TACAN station.

feet wet or dry Flying over water or land.

fence (in/out) Set cockpit switches as appropriate before entering or exiting the combat area.

fireball Possible multiple missile launches by Patriot being conducted (such as in response to a TBM attack).

flank/flanking Target with a stable aspect of 120 to 150 degrees.

flash Temporarily turn on prebriefed IFF mode.

float Directive or information to expand the formation laterally within visual limits to maintain radar contact or prepare for a defensive response.

fox (number) Simulated air-to-air weapons employment.

ONE - Semiactive radar-guided missile.

TWO - Infrared-guided missile.

THREE - Active radar-guided missile.

fox three close AIM-120 launched inside MPRF active range.

fox mike VHF/FM radio.

friendly A positively identified friendly contact.

fur ball A turning fight involving multiple aircraft.

gadget Radar or emitter equipment.

gimbal (direction) Radar target is approaching azimuth or elevation limits.

gingerbread Alert that voice imitative deception is suspected on this net.

go active Go to briefed Have Quick net.

goggle/degoggle Directive or information to put on or take off NVG.

gorilla Large force of indeterminable numbers and formation.

go secure Activate secure voice communications.

grand slam All hostile aircraft of a designated track (or against which a mission was tasked) are shot down.

green (direction) Direction determined to be clearest of enemy air-to-air activity.

greyhound Friendly ground attack cruise missile.

group Radar targets within approximately 3 nautical miles of each other.

guns An air-to-air or air-to-surface gunshot.

hard (direction) High-G, energy sustaining turn.

head/head on 1. Target with an aspect of 160 to 180 degrees.

2. (USN) Target with an aspect of 0 to 20 degrees.

heads up Alert of an activity of interest.

heavy A group known to contain three or more contacts.

high Between 25,000 and 40,000 feet MSL.

hit 1. (A/A) Radar return in search.

2. (A/G) Weapons impact within lethal distance.

hold down Directive to key transmitter for DF steer.

holding hands Aircraft in visual formation.

hold fire An emergency fire control order used to stop firing on a designated target, to include destruction of any missiles in flight.

home plate Home airfield or carrier.

hook (left/right) Directive to perform an in-place 180-degree turn.

hostile A contact positively identified as enemy according to theater rules of engagement; contact may be engaged.

hot 1. Attack geometry will result in roll-out in front of the target.

2. On a leg of the CAP pointing toward the anticipated threats.

3. Threat group heading toward fighters. Opposite of "cold.".

4. Ordnance employment intended or completed.

hot dog Information or directive call that an aircraft is approaching or at a specified stand-off distance from the sovereign airspace of a nation (as defined by national boundaries or territorial sea and airspace). (Color may indicate additional stand-off distance.) Follow briefed procedures.

hotel fox HF radio.

husky (USN) AIM-120 supported to HPRF active. Same as USAF cheap shot.

ID 1. Directive to intercept and identify the target.

2. ID accomplished; follow with type.

in (direction) Information indicating a turn to a hot aspect relative to a threat. Opposite of "out."

India IFF Mode IV.

interrogate Interrogate the designated contact of the IFF mode indicated.

jink Unpredictable maneuvers to negate a gun-tracking solution.

joker Fuel state above bingo at which separation, bugout, or event termination should begin.

judy Aircrew has radar or visual contact on the correct target, has taken control of the intercept, and requires only situation awareness information. Controller will minimize radio transmissions.

kill 1. Clearance to fire.

2. In training, a fighter call to indicated that kill criteria have been fulfilled.

knock it off Directive to cease air combat maneuvers, attacks, or exercise activities.

ladder Three or more groups or contacts in range.

laser on Directive to start laser designation.

lead/trail Tactical formation of two contacts within a group separated in range or following one another.

leaker(s) An aircraft that has passed through the A/A forces. Call should include amplifying information.

line abreast Two contacts within a group (side-by-side).

lights on/off Directive to turn on or off exterior lights.

locked Final radar lock-on; sort is not assumed.
(BRAA/direction)

low Target altitude is below 10,000 feet AGL.

mad dog Visual AIM-120 launch.

magnum Launch of friendly harm missile.

mapping Multifunction radar in an A/G mode.

marking Friendly aircraft leaving contrails.

marshal/ marshaling Established at a specific point.

medium Target altitude between 10,000 feet AGL and 25,000 feet MSL.

merge(d) 1. Information that friendlies and targets have arrived in the same visual arena.

2. Call indicating radar returns have come together.

mickey Have Quick time-of-day signal.

midnight Information advising that C^2 functions are no longer available. Opposite of "sunrise."

monitor Maintain radar awareness on or assume responsibility on specified group.

mother Parent ship.

mud (direction/type) Indicates unknown RWR ground threat displayed followed by clock position and type.

music Electronic deceptive radar jamming.

nails RWR indication of AI radar in search. Add clock position and azimuth, if known.

naked No RWR indications.

new picture Used by controller or aircrew when tactical picture has changed. Supersedes all previous calls and reestablishes picture for all players.

no factor Not a threat.

no joy Aircrew does not have visual contact with the target, bandit, or landmark; opposite of "tally."

notch (direction) All-aspect missile defensive maneuver to place threat radar or missile on the beam.

off (direction) Informative call indicating that the attack is terminated; maneuvering to the indicated direction.

offset (direction) Informative call indicating maneuver in a specified direction with reference to the target

on station Information that the unit or aircraft has reached assigned station.

opening Increasing in range.

out (direction) Information indicating a turn to a cold aspect relative to the threat. Opposite of in.

package Geographically isolated collection of groups, contacts, or formations.

padlocked Informative call indicating that the aircrew cannot take its eyes off an aircraft or ground target without risk of losing tally or visual.

paint blue Correct ATO, ATM, or other air warfare directives IFF interrogation response.

paint white Incorrect ATO, ATM, or other air warfare directives; IFF interrogation response.

parrot IFF transponder.

picture Situation briefing that includes real-time information pertinent to a specific mission.

pigeons (location) Magnetic bearing and range to a home plate or specified destination.

pince/pincer Threat maneuvering for a bracket attack.

pit bull 1. Information that the AIM-120 is at MPRF active range.

2. AIM-54 at active range.

playmate Cooperating aircraft.

play time Amount of time aircraft can remain on station.

pop Starting climb for air-to-surface attack.

popeye Flying in clouds or area of reduced visibility.

pop up Information that a contact has suddenly appeared.

posit Request for position; response in terms of a geographic landmark or off a common reference point.

post attack (direction) Directive transmission to indicate desired direction after completion of intercept or engagement.

post hole Rapid descending spiral.

press Directive to continue the attack; mutual support will be maintained. Supportive role will be assumed.

print (type) Valid NCTR reply.

pump A briefed maneuver to low aspect to stop closure on the threat or geographical boundary with the intent to re-engage.

pure Information indicating pure pursuit is being used or directive to go pure pursuit.

push (channel) Go to designated frequency.

pushing Departing designated point.

pushing (group direction) Information that said group(s) have turned cold and will continue to be monitored.

range Two groups separated in range.

ray gun 1. Indicates a radar lock-on to an unknown aircraft that is presumed

(position/ to be a threat.
heading/altitude) 2. A request for a "buddy spike" reply from friendly aircraft meeting these parameters.

reference Directive to assume stated heading.
(direction)

reset Proceed to a prebriefed position or area of operation.

resume Resume last formation, station, or mission ordered.

retrograde Directive to withdraw from the present position or area of operation in response to a threat.

rider A bogey that is conforming with safe passage routing, airspeed, or altitude procedures.

rifle AGM-65 maverick launch.

ripple Two or more munitions released or fired in close succession.

rolex (time) Time line adjustment in minutes from preplanned mission execution time.

rope Illumination of an aircraft with an IR pointer.

rumba On-ship maneuvering and ranging.

saddled Information from wingman or element indicating the return to the briefed formation position.

SAM (direction) Visual acquisition of a SAM or SAM launch (should include position).

sandwiched A situation where an aircraft or element is positioned between opposing aircraft or elements.

saunter Fly at best endurance.

scram Directive to proceed to a safe area for defensive or survival reasons.

scramble Take off as quickly as possible.

scud Any threat TBM.

separate Leave a specific engagement; may or may not reenter.

shackle One weave; a single crossing of flight paths; maneuver to adjust/ or regain formation parameters.

shadow Follow indicated target.

shift Directive to shift laser illumination from offset to target.

shooter Aircraft or unit designated to employ ordnance.

shot gun Prebriefed weapons state at which separation or bugout should begin.

skate Information or directive to execute launch and leave tactics.

skip it Veto of fighter commit; usually followed with further directions.

skosh Aircraft is out of or unable to employ active radar missiles.

skunk A surface contact that is unidentified but assumed to be the enemy.

slap shot Directive for an aircraft to employ a range-unknown HARM against a specified threat at the specified bearing.

slow A target with a ground speed of 300 knots or less.

smash A directive to turn on or off the anticollision lights.

smoke Smoke marker used to mark position.

snake A directive to oscillate an IR pointer about a target.

snap (direction) An immediate vector to the group described.

sniff Passive NCTR reply.

sniper A directive for an aircraft to employ a range-known HARM against a specified threat at the specified location.

snooze A directive or information indicating initiation of EMCON procedures.

sort Directive to assign responsibility within a group; criteria can be met visually, electronically (radar), or both.

sorted Sort responsibility has been met.

sour 1. Invalid or lack of IFF response.

2. Opposite of "sweet."

spades No IFF response.

sparkle Target marking by IR pointer. Target marketing by gunship or FAC-A using incendiary rounds.

spike RWR indication of an AI threat in track, launch, or unknown mode. Include bearing/clock position and threat type, if known.

spin A directive or information to execute a prebriefed timing or spacing maneuver.

spitter (direction) An aircraft that has departed from the engagement or is departing the engaged fighters targeting responsibility.

splash 1. Air target destroyed.

2. Weapons impact.

split 1. Request to engage a threat; visual may not be maintained; requires flight lead acknowledgment (air-to-air).

2. Directive to begin briefed maneuver or attack.

spoofing Information that voice deception is being employed.

spot The acquisition of laser designation.

squawk Operate IFF as indicated or IFF is operating as indicated.

stack Two or more groups, contacts, or formations with a high or low altitude separation in relation to each other.

status Request for a tactical situation or position.

steady A directive to stop oscillation of IR pointer.

steer Set magnetic heading indicated.

stern A request for or directive to intercept using stern geometry.

stinger A formation of two or more aircraft with a single aircraft in trail.

stop Stop IR illumination of a target.

stranger Unidentified traffic that is not a participant in the action in progress.

strangle ( ) Turn off the equipment indicated.

strip Individual fighter or section is leaving the formation to pursue separate attacks.

stripped Informative call from the wingman or element indicating out-of-briefed formation or position.

strobe Radar indications of noise jamming.

sunshine Informative C^2 functions are available; opposite of "midnight."

sunrise Informative command and control functions are available from GCI, AWACS, or AIC; opposite of "midnight."

sweet 1. Equipment indicated is operating efficiently.

2. Valid response to an administrative IFF check.

switch/switched Indicates an attacker is changing from one aircraft to another.

tactical A request or directive to switch to tactical control.

tally The sighting of a target, bandit, or landmark; opposite of "no joy."

target A directive to assign group responsibility to aircraft in a flight.

targeted ( ) Group responsibility has been met.

ten seconds A directive to the terminal controller to standby for the "laser on" call in approximately ten seconds.

terminate 1. Stop laser illumination of a target.

2. Cease local engagement without affecting the overall exercise.

threat (direction) Untargeted hostile, bandit, or bogey is within the prebriefed range or aspect of a friendly.

throttles Reduction in power to decrease IR signature.

tied Positive radar contact with element/aircraft.

tiger Enough fuel and ordnance to accept a commit.

timber Air control NPG Link 16/TADIL J.

tracking 1. Stabilized gun solution.

2. Continuous illumination of a target.

3. Contact heading.

trailer The last aircraft in a formation.

trashed Information that a missile has been defeated.

tumbleweed Indicates limited situational awareness; no tally, no visual; a request for information.

unable Cannot comply as requested or directed.

uniform UHF radio.

vampire Hostile antiship missile.

very high Above 40,000 feet MSL.

vic Three groups, contacts, or formations with the single closest in range and an element in trail.

victor VHF-AM radio.

visual Sighting of a friendly aircraft or ground position; opposite of "blind."

wall Three or more groups, contacts, or formations in line abreast.

warning (color) Hostile attack is--

RED -- Imminent or in progress.

YELLOW -- Probable.

WHITE -- Improbable (all clear).

weapons ( ) Fire only--

FREE -- at targets not identified as friendly according to current ROE.

TIGHT -- at targets positively identified as hostile according to current ROE.

SAFE -- in self-defense or in response to a formal order.

weeds Indicates that aircraft are operating below 2,000 feet AGL.

what luck Request for results of missions or tasks.

what state Report amount of fuel and missiles remaining. Ammunition and oxygen are reported only when specifically requested or critical.

( ) Active. The number of active radar missiles remaining

( ) Radar. The number of semiactive radar missiles number remaining.

( ) Heat. The number of IR missiles remaining.

( ) Fuel. Pounds of fuel or time remaining.

winchester No ordnance remaining.

words Mission-pertinent information.

working 1. A SEAD aircraft is gathering electronic order of battle on a designated emitter. Generally followed by signal type (SAM, AAA, or group) and bearing and range, if possible..

2. A/A aircraft executing EID on a specific aircraft or group to obtain identification necessary for BVR employment.

yardstick Directive to use A/A TACAN for ranging.

REFERENCES

DOCUMENTS NEEDED*

These documents must be available to the intended users of this publication.

Army Regulations

AR 95-1. Army Aviation: Flight Regulations, 30 May 1990.

AR 95-3. Aviation: General Provisions, Training, Standardization, and Resource Management, 27 September 1990.

Allied Communication Publication

ACP 165. (C/NOFORN)Operational Brevity Codes(U) (Allied Restricted), 1 March 1978.

Department of the Army Form

DA Form 2028. Recommended Changes to Publications and Blank Forms, 1 February 1974.

Department of Defense Forms

DD Form 175. Military Flight Plan, May 1986.

DD Form 1348. DOD Single Line Item Requisition System Document (Manual), July 1991.

Department of Defense Publications

DOD Dir 4000.19. Basic Policies and Principles for Interservices, Interdepartmental, and Interagency Support Manual, October 1994.

DOD Manual 4000.19. Defense Retail Interservices Support Manual, October 1994.

*These sources also were used to develop this publication.

Federal Aviation Administration Publication

FAR Part 91. General Operating and Flight Rules, March 1974.

This publication is available from Director, US Army Aeronautical Services Agency, ATTN: MOAS-AI, Cameron Station, Alexandria, VA 22304-5050.

Field Manual

FM 100-5. Operations, 14 June 1993.

Joint Services Publications

JP 3-04. Doctrine for Joint Maritime Operations (Air), 31 July 1991.

JP 3-04-1. Joint Tactics, Techniques, and Procedures for Shipboard Helicopter Operations, 28 June 1993.

US Navy Publications

NAEC-ENG-7576. Shipboard Aviation Facilities Resume, 1 January 1996.

NAVAIR 00-80R-14. US Navy Aircraft Firefighting and Rescue Manual, 15 March 1994.

NAVAIR 00-80R-19. US Navy Aircraft Crash and Salvage Operations Manual (Afloat), 14 March 1994.

NAVAIR 00-80R-20. US Navy Aircraft Crash and Salvage Operations Manual (Ashore), 15 March 1994.

NAVAIR 00-80T-105. CV NATOPS Manual, 1 November 1995.

NAVAIR 00-80T-106. LHA and LPH NATOPS Manual, 1 August 1994.

NAVAIR 00-80T-113. Aircraft Signal NATOPS Manual, 1 June 1988.

NAVAIR 00-80T-113(Change 1). Aircraft Signal NATOPS Manual, 15 July 1992.

NAVSEA OP-3565. Electromagnetic Radiation Hazards (Hazards to Personnel), Fuel, and Other Flammable Materials, 15 July 1989.

NAVSEA OP-4. Ammunition Afloat, 15 February 1972.

OPNAVINST 3721. Air Traffic Control Facilities Manual, 1 March 1984.

OPNAVINST 3750.6-Series. Naval Aviation Safety Program, 28 August 1989.

The US Navy publications listed above are available from the Commanding Officer, Naval Air Technical Services Facility, Code 25, 700 Robbins Avenue, Philadelphia, PA 9111-5097, or by calling the Navy Publications Customer Service Center, Philadelphia, Pennsylvania, at DSN 442-2997/2626/0160/2267.

NWP 3-02.1. Ship-to-Shore Movement, 1 August 1993.

NWP 3-04.1. Shipboard Helicopter Operating Procedures, expected publication date: 3rd quarter FY 97.

The US Navy publications listed above are available through the Commander, Naval Doctrine Command, 1540 Gilbert Street, Norfolk, VA 23511-2785, to the Director, Navy Tactical Support Activity, 901 M Street SE, Building 200 (Code 57), Washington, DC 20374-5079, or by calling the Navy Tactical Support Activity at DSN 228-6163.

Made in the USA
Monee, IL
04 November 2026

67387200R00154